HOW TO
DISAPPEAR AND
NEVER BE FOUND

HOW TO DISAPPEAR AND NEVER BE FOUND

BARRY DAVIES, BEM

Skyhorse Publishing

Skyhorse Publishing books may be purchased in bulk at special discounts for sales promotion, corporate gifts, fund-raising, or educational purposes. Special editions can also be created to specifications. For details, contact the Special Sales Department, Skyhorse Publishing, 307 West 36th Street, 11th Floor, New York, NY 10018 or info@skyhorsepublishing.com.

Skyhorse® and Skyhorse Publishing® are registered trademarks of Skyhorse Publishing, Inc.®, a Delaware corporation.

Visit our website at www.skyhorsepublishing.com.

10 9 8 7 6 5

Library of Congress Cataloging-in-Publication Data is available on file.

ISBN: 978-1-5107-5267-2
Ebook ISBN: 978-1-62636-521-6

Cover design by Daniel Brount
Cover photographs courtesy of Getty Images

Printed in China

To Andrew Rhys Howell

When I first met Andy it was like jumping into wet concrete—the longer I stayed with him, the harder it was to leave. When I do manage to finally escape there will be permanent footprints left behind.

TABLE OF CONTENTS »

INTRODUCTION »

The title of this book is a clear indication of its content: a guide for those who wish to disappear off the face of the earth and never be found. It sounds simple, but as the world and society mature, it becomes increasingly difficult. Disappearing and starting a new life is achievable, but it takes meticulous planning, nerves of steel, and a promise that you will never look back.

There are several aspects that will affect your ability to disappear. For example, your loved ones—can you accept leaving them behind and never seeing them again? If you are famous, your face will give you away, and if you are rich and continue to live in luxury after your disappearance, you will easily be found.

These are just a few of the factors that will govern your disappearance. In the end, your success comes down to you and your individual strength to hold your plan together, stick to the rules, and stay below the radar.

You will need to know what your new life is going to be and where it will be located. Don't always think that the grass is greener on the other side, because more often than not, it will disappoint you and you'll find yourself worse off than before. You will need to make a plan that is foolproof, prepare a year or more in advance, and lay down clear and sustainable disinformation. Put your safeguards and new life support program in place, test them, build your cover story, and become a different person the moment you leave your old life behind.

* * *

The essence of man is to survive and live in a condition that is relaxed and tolerable. When these are not present, we look to adjust our lives so that life is, at worst, sufferable. Nevertheless, for some of us there comes a time when we cannot modify our lives, and the only option open to us is to change them and start anew. The will to live another life is not in all of us, but burns brightly in a handful of people. When man is against the odds and struggles to overcome adverse circumstances, dramatic change is often the only way out: in effect, to disappear. In order to be successful, you must act alone, seeking no help or advice from others. Until you are ready, you must continue to suffer the humiliations, trials, and tribulations that forced you into action in the first place. You must plan, gauge your resources and any elements you have at your disposal, make a plan, and rely solely on your own worth until the day you disappear. With the right planning and cover story, matched by good security and adequate funding, you can disappear and never be found again.

* * *

In writing this book, I have taken into account two reasons for you reading this book:

1. That you intend to plan in detail your disappearance
2. That you simply walk out on life and become a hobo.

The first reason takes up the majority of the book and highlights a disappearance plan and all that it entails. Living like a hobo is more of a survival adventure, but I hasten to add that this is not a survival book, and I have limited this information to the bare bones. That said, I am a military man by nature and hence my thinking and logic is military. Thus, much of this book is based on my life's knowledge and experience. In as much as I have traveled this planet from corner to corner—and continue to do so—we live in a world where hiding from society is not a matter of luck, but of detailed planning.

Should you ever use this book and really try to disappear, then I wish you luck. But as the old Japanese proverb says, "to wait for luck is the same as waiting for death."

WHY WOULD YOU WANT TO DISAPPEAR?

The first question I would have to ask is why anyone would want to disappear? Well that's just me: I like my life and I am happy, but there are many people in this world who would rather die than carry on the way they're living. So there are countless reasons why someone disappears; it might be planned, accidental, or involuntary. In most cases when a person disappears, there are extenuating circumstances that indicate what kind of disappearance it is. The list of explanations is long, and below are just a few reasons why people would want to disappear:

- Many young children disappear because they wish to escape from child abuse, either emotional or physical (they usually end up being abused further).
- Teenagers may disappear because they fall in love with someone their family disapproves of, running away to live with their partner.

- Others—for a number of reasons—may wish to disappear and quietly commit suicide rather than carry on living.
- Many adventurers have disappeared, mainly due to accidents in the wild or at sea.
- Others plan to rid themselves of an unruly spouse or for financial gain, such as insurance fraud.
- Many individual travelers are robbed, raped, murdered, and their bodies disposed of.
- There is a huge market in human trafficking; people are sold into slavery or a life of prostitution. Most end up dead from a drug overdose or lying in a ditch with their head removed for disobedience.
- Some people disappear simply because the government wishes it—this might be a political rival or drugs baron with vast control over a specific region.

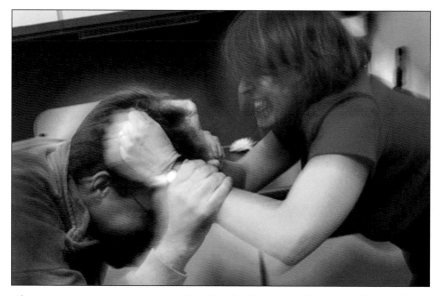

Abuse is a common occurrence with either husband or wife being to blame. Often, this creates the desire to walk out of the relationship.

The list is perhaps endless and the reasons are varied. Just because a person disappears, it should not be automatically assumed that they are dead. In many cases, missing persons do eventually return home, or their whereabouts become known. On the other hand, in certain circumstances it can be accepted that the person is actually dead. For example, some 2,600 American servicemen were listed as missing in action during the Vietnam War. These servicemen have a legitimate reason that explains their disappearance. While it is true that some might have been captured and held prisoner for many years, the majority were probably killed in a firefight and left behind to rot in the jungles of Vietnam. Their bodies were never discovered by anyone who had the time or inclination to find out the identity of the corpse.

A better known disappearance of an individual is that of aviator Amelia Earhart, who disappeared while flying over the Pacific in 1937. No wreckage was ever found and there was speculation that she had merely chosen to vanish. However, with her high public profile, this would have been virtually impossible for her to achieve. Given the circumstances and dangers inherent in her trip, it is far more reasonable to assume that she simply crashed into the sea and died.

Lord Lucan simply disappeared.

The point I am trying to make is that when a person disappears, circumstances around the disappearance, the person's state of mind, and recent actions can contribute to legal presumption that the person has actually died. Where there are no circumstances to suggest a person is dead but that they have simply disappeared, then it is for the courts to rule on when that person is considered legally dead. This ruling will affect any estate the person has left behind, as well as any insurance claims. Therefore, while making your disappearance plan, you must factor in all the possibilities that could indicate you are simply missing and not dead, should you wish to leave any financial benefits such as pensions or insurance plans to your loved ones. Moreover, you must do this in such a way as not to arouse suspicion.

* * *

Before we continue, I must stress that it is not illegal to disappear unless you are trying to avoid debt, have done something unlawful, or if you intend to defraud an insurance company, etc. The simple fact is that people go missing every day: In the United States alone, an astounding 2,300 Americans are reported missing every day, including both adults and children. As mentioned, in many cases there are circumstances surrounding the disappearance of a person that point to the reason he or she is missing. In addition to circumstance, there are other logical reasons for people to have disappeared, such as that they have committed murder, stolen money, accumulated large debts, and so on. For the majority of these people, their disappearance is controlled by the circumstances they find themselves in and it is not something they had time to plan in advance.

This book is dedicated to providing information for those who simply want to disappear and never be found again. In doing so, I have avoided the usual disappearing material found in most books on the subject and utilized my military background and the fact that I have traveled much of our wonderful planet. If it is your intention to completely disappear and never be found, then you need to make a proper plan. However, before you do this, there is one important consideration that most of us will have to face: leaving our loved ones behind. This has to be the toughest of all the decisions that need to be made; to leave one's family, husband or wife, children, friends, mother and father, and never see them again.

No matter the reason for leaving, the thought of your loved ones will always be with you. Try to think how they will feel if you disappeared, never to be seen again. You may well have left some provisions to cover living expenses, but think of a young mother with children and her thoughts on how she is going to survive without you. Or a father who finds his wife has disappeared and is left with the children. How will they cope with going to work and looking after the children, especially if they're very young?

Leaving loved ones behind must be the most difficult part of disappearing.

You must also consider when you disappear that your family will not know what actually happened to you. If your plan is being carried out with attention to detail and they have no suspicion of your intentions, then they will not understand the reason for your disappearance. They may think you've been killed in an accident, run off with another person, or even that you've been kidnapped. The point being, they will be left not knowing if you are alive or dead. Just think how many people your disappearance will affect: your direct family, your close friends, the insurance company, the police, the missing persons bureau—all of them will be looking to find you or confirm the means of your demise.

Before making any plan to disappear, it is better to first consider whether it is possible to change the circumstances that are driving you away. See whether it is possible to continue your present life and change it for the better. Change is always possible; victims of violence and people with heavy debts can always find help and guidance to make their present lives bearable. No matter what trouble you are in, there are many organizations—both in the United Kingdom and United States—that will help put your life together again or to put your problems in perspective.

At some time in our lives, we will all come face to face with a major problem that might not be of our own choosing and might even be forced upon us, but face it we must, for the alternative is not worth contemplating. Life, as precious as it is, can sometimes feel subordinate to death. So great is the feeling of helplessness that suicide comes to the forefront as an option to end the pain. Thankfully, there is always light at the end of the tunnel and things do work themselves out.

Author's Note: In 1990, I separated from my wife after seventeen years of marriage. The fact that I loved my wife and freely admit that her leaving was due to my own adultery does not make the situation any easier. The children—my son, aged sixteen, and my daughter, aged fourteen—decided to stay with me and not go with their mother. So there I was, a single parent with no job and about to start work on building a new home, which we had planned for our family.

The first few days I just walked around in disbelief, while the children seem to bear their mother's departure a lot better than I did it was the start of a horrendous six months. After about a week, I felt that my whole world had collapsed around me, and the hole in my chest seemed to get bigger day by day. It is difficult to describe those feelings to anyone other than those who have suffered a similar situation. Luckily, I had two children to look after and our new house to build, which occupied my mind. My son helped me with the house construction and my daughter continued to go to school. In the evenings

I continued with my writing, hoping to finish my first book. If I'm honest, I really enjoyed looking after my children, as previously I spent most of my life away on military operations. After about three months, I started to get a feeling that there was something else out there pushing me toward a new purpose. Initially I thought I was going to win the National Lottery, and the feeling grew stronger by the day. What I didn't realize is that it was being generated by my writing, something I had taken up each evening to alleviate my broken heart. At long last, after I had completed three books, I decided to send one to a publisher.

I had previously purchased a book titled *How to Get Yourself Published* and while this was a great read and offered lots of advice, it was difficult to see how it would get me into the world of writing. Then came the biggest shock of all: three days after I sent my first manuscript to Bloomsbury Publishing (the same people who publish *Harry Potter* in the UK), I received a telephone call from the managing director asking to see me. While on the telephone, I told him about the other two books and he asked me to bring them all up to London. Within less than two weeks I had a contract and a check for all three books, which set me on the path of my writing career.

The reason I tell you this is simple. No matter how bad your situation or your circumstances might seem, you never know what's around the corner. For me, the feeling that was so strong inside me pushed me to become a writer; something I enjoy to this day that turned my whole life around. Actor Ben Affleck recently summed this up when he stated, "It doesn't matter how you get knocked down in life, all that matters is that you get up."

Do not try to fake your own death or pretend to have committed suicide by leaving your clothes on the seashore or disappearing from a ship. Leaving your personal effects behind always raises some suspicion of doubt, especially if there is a large insurance payout. There are cases where people have tried to disappear using someone else's death—disfigured bodies caused by various accidents such as fire or long-term drowning at sea—to cover their tracks. I can assure you that this is not the path to take, as the true identity of any dead victim will soon be established and your disappearance discovered.

Faking your own death to collect life insurance is more common than we think, with thousands of people trying to outwit insurance companies every day. There are people who have killed people who look similar to themselves and claim the body is theirs. One guy even tried to take advantage of an insurance company when the Twin Towers were brought down by terrorists, claiming that his wife had been in one of the buildings.

Some people have gone to the extent of using a disfigured body as part of an elaborate insurance scam. Few succeed in fooling modern day forensics.

Unless you have a 100 percent waterproof plan to defraud insurance companies, I would suggest you don't do it. No doubt there have been many successful cases where people have staged their own death and claimed the insurance. But if they exist we will never know for as far a society is concerned they are dead—or are known as someone else.

* * *

Most people who run away from a serious crime and start another life in a different country usually turn to crime again, and their life becomes the same as it was before. More likely than not, their crimes will catch up with them and they will either be caught or will be on the run once more. Criminals will always be criminals; those who pursue a path of violence and easy money will never change. If you made a living selling drugs in America and then fled justice to live in Thailand, I would lay money down that sooner or later you will end up selling drugs in Thailand. If you get caught, you are in for a serious shock (believe me, the Bangkok Hilton is not a hotel), and the death penalty or life in jail is not a promising prospect.

Therefore, I do not intend to understand the reason why anyone would want to disappear and start a new life; I'm simply going to show you how to do it. I take no responsibility for what I write here; it's just information I have gathered over the years, mainly from my SAS military service and travel experiences. I have spent much of my time in surveillance—both in urban and rural conditions—and you would be surprised at the many places I've found to hide for weeks or even months at a time. It is true that I had a support unit to provide food and water so I could remain at the observation point without being discovered, but the principles of hiding and not being discovered remain the same.

As mentioned above, people all over the world will find a reason why they would want to disappear. In addition to many of the obvious reasons, such as murder, debt, rape, depression, and so on, there are also many other reasons for leaving your identity behind. The world is full of places controlled by dictators, unrighteous regimes, police states, and governments practicing religious and sexual oppression. Citizens from these places would happily disappear just to rid themselves of the daily drudge under tyranny. For them, disappearing would mean freedom, even if it means trekking miles with crying children only to spend several years in a refugee camp—millions of people find themselves forced into this situation every year.

Sometimes people disappear and hardly anyone notices; these people are loners, divorced men or women who have already parted from their families, or people who have simply been working overseas for so many years that even close family have them only on the back–burner of their thoughts. For these people, it is simply a matter of clearing their public records as much as possible. Once again, if they have been working overseas for many years, this will be minimal.

In a few cases, people disappear through no fault of their own. Young people—especially girls—are kidnapped and shipped off to faraway places where they are used for sexual exploitation. Furthermore, their disappearance seems to be magical, with all traces of them wiped away. Despite massive media coverage, public awareness, rewards, and substantial police investigations, many young girls are simply never seen again. Madeleine McCann was just one instance when she disappeared while on vacation with her parents in Portugal.

Madeleine McCann disappeared right under her parents' noses.

Madeleine McCann disappeared on the evening of Thursday, May 3, 2007 from the first–floor apartment where her family had been staying on vacation in Portugal. Madeleine's parents, Kate and Gerry McCann, told police they left the children in a ground–floor bedroom while they took their evening meal in a restaurant about 120 meters (400 feet) away. When the parents returned, they found that Madeleine was gone and immediately called the police. Despite a worldwide appeal and a reward of £2.5 million, not one single piece of factual information that would lead to how or why she disappeared has ever come to light.

Disfigurement Disappearance

I mention this here because it's one of the "tradecraft" tools used by various government agencies to remove the identity of someone they have killed, and the person simply becomes a "John Doe." This tradecraft is a skill used by a "clean-up" or "wet" team, whose job it is to make sure the body cannot be identified. This entails removing or disfiguring all body parts that might be on record, such as fingerprints, iris and the retina, teeth, or distinguishing body marks, such as tattoos and full facial scans. One or more of these potential leads could allow the public authorities to identify the dead body. Total disfigurement is the only way to overcome these identity biometrics. The most common method used is acid. If a strong solution of sulphuric acid is poured over the hands and face, it will totally eradicate the facial structure, including the eyes and enamel from the teeth, while also burning off fingerprints. An alternative to acid is removing the teeth with a hammer and dropping the body in the sea. Long-term (two weeks) exposure to marine life will certainly do the trick. Personally, I think it's an awful lot of wet work to arrange and you must ask yourself if it's really necessary. In many cases, government agents have not been so fussy and mass graves have been the norm for disposal of unwanted bodies.

Disappearance Services

There are many people out there who will offer to help you disappear, either by design or simply by selling you a how-to book. First of all, if you plan to disappear, never trust anyone other than yourself; never take written advice as gospel unless you have tested the methods personally.

It is possible to find agencies that will help you disappear in a way similar to how a witness protection program works. You are taken away to a safe house and protected until you are required to give testimony, after which point you are furnished with a new identity. The only thing I can say about individuals and companies offering services that will guarantee your disappearance is to be very careful. Most stipulate that you must travel to a foreign country and visit a specific address; you should not tell anyone you are going to disappear and neither should you say where you are going. Now how dumb do you have to be to actually do this?

Say for example, the company wishes you to travel to Bolivia and visit the town of Vallegrande; you should bring $10,000 with you, after which they guarantee you will disappear. Trust me, you will disappear and never be seen again. The trouble is that you will end up dead and buried in a hole someplace. So please don't bother answering any advertisements that guarantee your disappearance as they will probably honor the guarantee.

Barry Davies

Disappearance services might be genuine, but would you want to take the chance?

Understanding Your Lifestyle, Age, Ability, and Skills

Before you even think about disappearing and forming a new life, there are a few factors that will automatically govern your decision. Your lifestyle, age, ability, and skills will all affect any plans to disappear. As humans, we are all molded into various lifestyles. For example, you might work in the stock exchange and earn a great deal of money, or on the other hand, you might be a clerk working in a supermarket. This lifestyle mold is extremely hard to break, and while many believe they can always do better, it is a fact of life that we use the skills we have learned. Therefore, if you are a high-flyer in your old life, you will have a tendency to be a high-flyer all your life. If you were a vehicle mechanic in your past, then you will certainly look for similar work in your new life—it's just using the skills you have.

For example, it would be almost impossible for me to disappear completely and never be found. The reason being that I've written more than twenty-three books, most of which are for sale on the Internet. These I cannot retract or remove, and so I am doomed to always be Barry Davies. However, few people know what I presently look like, so I could change my name, get a new passport and driver's license, and travel to a new country to start a new life. While all these things are possible, I would not be able to leave my family behind. It would be the same for anyone who has a lifestyle that is known to many others (to prove a point, I have managed to remove all but one image of me from the Internet, however, there remain hundreds of videos on YouTube).

Age

Your age is also a vital factor when considering this. The majority of people who seriously consider this are generally between the ages of twenty-five and fifty. Many young people leave home because they've been abused by a family member. You can find these people on the streets of New York or London every day. They are wrapped in blankets, sleeping in shop doorways, and begging for money. While these people have run away, they have not disappeared; in addition, they are open to a whole range of abuses.

There also comes a time when you are just too old to disappear. If you have reached the good old age of, say fifty-five, it's going to be hard, and by the age of sixty you will most likely be accustomed to your lifestyle. As we grow older, our bodies become frail and we might need medication or other state benefits, so disappearing becomes less of an option. No matter how strong your determination, motivation, and confidence may be, the body ages and there is a point at which the aging process will simply deny any further progress. As we get older we lose muscle weight, flexibility, and our bones become more brittle. At what stage this occurs differs in all of us and in most cases is dictated by our lifestyle. For some it will come early, while others can remain active into their eighties.

If you try to disappear when you are in your teens, you will have very little worldly experience, and you will be of a sexually attractive age and therefore very vulnerable. If life is so bad that you need to run away as a teenager, the best advice I can give is to move in with a relative or friend you can trust. If you have no one, then go to the authorities.

Between twenty and forty years old, you will be at your peak. Confidence and motivation will be at its highest. You will understand a more about the world and what makes it tick. This is the best age bracket in which to disappear.

If you're forty and beyond then, you will need to look at what you have and what you are leaving behind. At forty or older, most of us are married, have children, have built a home, and maybe saved a little money for a rainy day. While it is still possible for you to disappear, my advice would be to first search for an alternative.

Beyond sixty, you have really waited too long to disappear. You may still have the drive and determination, but your body will start letting you down. Imagine if you decided to leave home one day and become a hobo—a year of being on the road will age you so fast you will end up looking ninety.

Ability and Skills

A very important requirement for being able to disappear must be the ability and skills you possess. For example, a soldier who has seen action around the world who is in his early thirties with no physical disabilities would have no problem disappearing just about anywhere on the planet. By comparison, a young mother of similar age, with two or more small children, would find it hard to even contemplate leaving her family behind.

Skills are something we have learned; these might be physical, such as being a mechanic, or practical, such as being a housewife working around the home—but they are all skills. A mechanic can find work in most countries and a housewife could easily turn her hand to working as a maid in a hotel. In each case, we understand how to do something. Most people have a lot of skills: some speak a foreign language, another may be proficient in a specialized subject such as sky diving or skiing, while others are professionals, such as doctors or soldiers. Anything you learn is a skill, and skills are crucial when it comes to disappearing. Skills give you confidence, can earn you money, and help you steal a car or pick someone's pocket.

I have added a list of skills later in this book that might be worth learning if you intend to disappear. In their own context, none are illegal unless you use them for illegal purposes. These skills are intended more for the hobo section (Chapters Eight and Nine), whereby you will be living on your wits, cunning, and ability, rather than financial support.

Health

You should also take into consideration your health. While you might be in perfect shape when you actually disappear, there is always the possibility that you will fall ill, contract some contagious disease, or be in an accident. If you are in your fifties, you might be on medication or showing signs of an age-related illness. In any case, you should always check what the medical facilities are like in any country you think of disappearing to. In addition, you should consider the cost if you are not a resident—hospital fees can be very expensive.

Author's Note: A friend of mine, let's call him George, went to Thailand and remains there to this day, happily married. What you don't know is that in his first year there, George spent most of his money. Although he had found employment as a teacher almost right away, the appeal of the nightlife in Bangkok led him to go out most weekends, staying in hotels and over-indulging on alcohol and women (this happens a lot to those who have never been to Thailand before).

Barry Davies

George and his wife Whaw are happy despite the fact that George speaks little Thai and Whaw very little English.

Some eight months in, George purchased some locally made Viagra in the hope of increasing his waning love life. Sure enough, George took three tablets and was all set for a fantastic weekend. Some hours later, while lying in a hotel bedroom next to a beautiful young Thai woman, George's chest suddenly gripped him like a vice. He had the good sense to get the girl out of there, jump in a taxi, and make his way home. But on the way home, the pain was so severe he asked the taxi driver to divert to the nearest hospital. This was perhaps the luckiest move George ever made, as had he not decided to leave the hotel, he probably would have died.

He told the doctors the truth and had immediate heart surgery with several stents implanted in his heart. He needed weeks off from work to recover, which nearly cost him his job. Moreover, he had to pay out the equivalent of $15,000, which put a huge hole in his disappearance fund.

The lesson here is not to be tempted into anything that could impair your health, deplete your financial reserves, or put your earning potential in jeopardy. George did learn from his mistakes, and took to providing private English lessons to supplement his school income. Today, this enables him to live a quiet and happy life with his beautiful wife (there will be more about George later).

Short-Term Disappearance

In many cases, all some people need is a short-term disappearance to get their affairs in order. This disappearance might buy you time to collect enough money to pay off your debts or provide some space for you to reflect on your current situation should you be the victim of abuse. Many young teenagers run away because they do not get along with someone in the family home (or they are being abused). Most of these disappearances are short term because they do not really want to vanish; they simply want an end to the abuse or to change their living situation.

Disappearing for a short time is relatively easy. In most cases, people will simply stay with a close friend or distant relative. For those people who do not have either friends or family they can trust, it is always possible to find some location where they can hide and never be discovered. The problem is that most of those who disappear in the short term are young children or teenagers, and, without sounding detrimental to their generation, they give little consideration to where they will go or what they will do when they get there. Most of these young people flee without a thought; they have only a small amount of money for food and drinks, and at best a sleeping bag. I can offer little advice to these people other than to find a safe place after dark. Get away from the streets, get somewhere dry, and snuggle up in your sleeping bag or anything that will keep you warm. Come out onto the streets only during daylight hours. If you must beg for money, do so in a crowded area. Be aware that it's not the general public and police you need to watch out for, but others living rough or those that want to exploit you.

If you are young and have run away from home for reasons known only to you, never be afraid to go to the police or call a close friend. Once reported, you will be relatively safe or placed in proper care. Being a young girl or boy on the streets is not the best of situations, so before you decide to leave home, please read the horrendous stories that have happened to others close to your age. There is always an alternative and help available.

Many teenagers run away from home and sleep rough when they disappear from home. Life on the streets for young boys and girls is no picnic.

Military Short-Term Hide

While not directly related to disappearing completely, there is much to be learned from the methods used by Special Forces. The

British SAS have long been revered for their ability to hide or enter undetected. This is due to a long history of constructing hides that have enabled the SAS to stay concealed for weeks at a time while observing their enemy. The first of these is called an observation post, or "OP" in military terms. The second is referred to as a "hide," which is designed for staying behind enemy lines so they can attack the enemy in the rear. Years of painstaking training and refinement have created a set of simple rules for both the OP and the hide so they are never discovered.

Observation Post

An OP is a covert site from where enemy activity can be watched and intelligence gathered. The SAS are experts in setting up OPs and remaining in them for long stretches of time in the most hostile conditions. Sometimes they are in a rural situation and are constructed from natural materials to blend in with the surroundings. At other times they may be placed in an urban area, such as the loft space of a house or underneath a garden shed.

Wherever they are located, the rules for their construction remain the same. A site must be chosen that is not vulnerable to discovery and yet must also afford a good view of the target position. A concealed entrance and exit are also needed. High ground, although good for visibility, is an obvious spot and one that the enemy will search, so a less likely place is a better choice. Once the site has been chosen, the OP can be constructed under cover of night. It can be made out of any material: waterproof sheeting, ponchos, camouflage nets, and natural or locally available materials are all useful, as long as the end product blends in with everything around it and can't be easily seen. OPs tend to be built either in a rectangular shape, where the patrol members lie in two pairs facing opposite directions, or in a star shape where each member takes up an arm of the star.

As OPs are often maintained for long periods at a time—during the Falklands War, one OP was maintained for twenty-six days—so they have to contain all that is necessary for the inhabitants to be self-sufficient, particularly if it is not possible for them to be resupplied. In addition to food, water, clothing, and sleeping bags, operational gear is also stored inside the OP: weapons, radio equipment, binoculars, night sights, cameras, and telescopes. This can make conditions cramped and uncomfortable; a situation often made worse by weather conditions. No sign of the men's presence can be left since it may be discovered by an enemy. Therefore, even normally private functions such as urinating and defecating must be done in the OP into separate bags which are then sealed and taken away by the patrol members at the end of their operation. Other things that might not be appropriate in such a situation are smoking, cooking, and the wearing of deodorant or aftershave.

When living in such close conditions, it is essential that the men and women get along well. It also requires mental strength and the ability to get on with the job no matter how hard the conditions or how boring it might seem at times. However, these qualities are second nature to SAS soldiers.

A mock-up of an underground Operational Base at Parham Museum, Suffolk.

© British Resistance Archive

MEXE Shelter

Although now obsolete, the MEXE shelter was used by the two Territorial (Reserve) Army units—21 SAS and 23 SAS—when they were deployed in the forward observation, or "hide role," in Germany during the days of the Cold War. Designed to accommodate a patrol of four fully equipped men for a period of several weeks, it was installed underground in an excavation dug with the soil and turf being replaced thereafter to provide protection and camouflage. The main components were a steel frame, which had a load–bearing limit sufficient to support a vehicle crossing the ground above, and a prefabricated hatch unit, which provided entry to the shelter and a special skin. Once assembled, the frame was covered with the skin manufactured from a special composite fabric which was both waterproof and NBC agent (Nuclear, Biological and Chemical) proof. The latter property also kept scent from permeating through to the surface of the ground, thus preventing tracker dogs from detecting the presence of the occupants.

Many of these hides exist to this day and have never been removed. I once found one in a forest in Southern Germany (I had been given its exact location), and after a great deal of searching, I managed to find the trap door. The whole area was so overgrown that the trapdoor blended in perfectly with the surrounding vegetation. Inside, the shelter was still dry and could have been used again if need be. The communications aerial ran up out of the hide, under the ground, and up the side of a tree, which took me a long time to locate. This particular hide was not designed for observation but as a stay-behind hide—in the event of the former Soviet Union and its allies overrunning Southern Germany during the Cold War. The idea was that the hidden SAS men would emerge and attack the Russian command centers from the rear. This was not a new idea, as during the early days of WWII, thousands of young British troops were hidden in caves and mines, sealed in with weapons, food, and ammunition. Churchill set up the British Resistance Organization (BRO) with the idea to hit the Germans in the rear should they manage to successfully invade the British Isles. Fortunately, this never happened as the RAF broke the back of the Luftwaffe and the invasion never took place. Once it was realized that it was safe, the caves and mines were unsealed and the men and weapons used to reform a new army, one which eventually invaded and defeated Hitler.

Just once in a while we do get to know what actually happened to people who have disappeared. While it is interesting to learn that the British Special Forces are able to stay in underground bunkers (MEXE Shelters) for weeks at a time, you may think this has little or no relevance in a book on How to Disappear. You would be wrong. If I had to disappear for a while, digging a hole and living in it is a perfect way. However, I would need to make sure my hole were sufficient to support life. The former Iraqi leader Saddam Hussain was found in a concealed hole and only discovered due to the massive bounty on his head. But there is a far better portrayal of living underground and undiscovered, not for a few weeks but for twenty-four years. Unfortunately, this story is true and ranks as one of the most revolting crimes ever committed in recorded history.

Elisabeth Fritzl was abused by her father when she was just eleven years old. By the time she was eighteen, he had incarcerated her below ground. For twenty-four years, Josef Fritzl kept his daughter, and three of the seven children he incestuously fathered with her, locked away in a secret maze of underground rooms he'd built below his Austrian home. Josef Fritzl told his wife that their daughter had run off to join a cult, and for the next quarter of a century, he led two lives: one upstairs with his wife and one below the family home where he repeatedly raped his daughter. Elisabeth endured the rapes and subsequent births for twenty-four years before being discovered. Had one of the children not been seriously ill, she may have lived out her life in her underground tomb.

Of the seven children, one died at birth, three remained with Elisabeth, and three were taken upstairs by *Josef Fritzl hid and abused his daughter below ground for more than twenty-four years.*

Josef Fritzl. His explanation for the children's appearance was that they had been left on his property—no one seemed to question the periodical appearance of children in the family home, not even his wife. The remaining three, a girl, aged nineteen, and two boys, aged eighteen and five, remained with their mother, buried alive, within the confines of a small apartment.

It was April 19, 2008, before Josef Fritzl's secret was discovered. Their eldest daughter became ill and Elisabeth somehow managed to persuade her father to take the girl to the hospital. The authorities became suspicious when Fritzl produced a letter supposedly written by his daughter. It looked false and provoked suspicion, leading to the case of missing Elisabeth being reopened. The report of her disappearance was televised and Elisabeth actually watched it in her underground cell. The next time her father visited her, she pleaded with him to let her go to the hospital. On April 26, he allowed Elisabeth and the other two children upstairs, explaining to her mother that she had decided to come home after all these years. She then went with her father to the hospital to see her daughter, at which time the doctors tipped off the police and both were held for questioning. Elisabeth would say nothing to the police until they promised her that she would never have to see her father again. Then she told them of her twenty-four years of captivity.

Although they suffered massive psychological and medical problems from their captivity, Elisabeth and the three children survived. If nothing else, this horrific story goes to prove that humans, given the right conditions, can disappear for years.

Self-Exile

Some people, for reasons known only to them, go into voluntary exile. Some do so because they think others are after them and mean to do them harm; others do it just for a change of lifestyle. Self-exile is the same as disappearing, but without trying to cover your tracks.

In the United Kingdom, a middle-aged man named Philip Sessarego pretended that he had been a member of the British Special Forces (SAS). He wrote a book about SAS undercover operations in Afghanistan using the pen name Tom Carew. In fact, he had never been a member of the SAS (although he had been a failed candidate of selection) or anywhere near Afghanistan, and the whole book was a complete piece of fiction. Even so, some of what he had written was extremely close to the truth with regards to SAS operations, because Sessarego had read and done research into the way the British SAS operated. His book, titled *Jihad*, gave him two years of fame before it was exposed as a hoax. The reason it took so long to uncover Sessarego as a fantasist was that the SAS says very little and never comments on any of its operations.

Convinced he would face assassination for exposing some of the techniques used by the British Special Forces, he disappeared to the Belgian city of Antwerp. Here he rented a lock-up garage and, using survival techniques that he formerly claimed to have mastered in the SAS, simply disappeared. His decomposing body was found by the landlord some months after he had died. The lock-up contained jerry cans of water, a small stove, and some basic sleep equipment. There was also a loaded pistol by his side. There were no signs of foul play, and it is believed he died by inhaling carbon monoxide fumes from his stove. Despite this, there was much speculation about his decomposing body, which was unrecognizable. Some of his family members believe that the body was not his and that it was just another elaborate plan to fake his own death. DNA samples were taken from family members in order to finally establish the identity of the body as his. His death left few people grieving as he was a man who made a lot of enemies and few friends.

"I wouldn't care if somebody killed him, because he brought it all on himself," said Claire, his daughter, in a recent interview, adding that there were "a lot of former SAS men" who had scores to settle with him and would be happy to have murdered him. So what had the soldier done to merit such hatred? Simple: He was regarded

Philip Sessarego played at being British Special Forces, but he died a sad and lonely death.

as a traitor and an "SAS wannabe" who had failed The Regiment's selection process, yet claimed throughout his military career that he was a member of the legendary British military elite.

George (mentioned earlier in this chapter) is another example of self-exile. He got divorced in the UK, lost his job, and ended up as a truck driver. Once a month or so I would meet him for a drink and he was so melancholy about his lost love (ex-wife), that it was painful to be with him. In the end, I took the initiative and told him to snap out of it and go to Thailand for a vacation. Some months later, he did. He then returned home to the UK, only to go back to Thailand the next month and then the month after that (Thailand has that effect on sorry souls).

The next time I met George, he was full of beans and told me of his plan: He would take a TEFL course and learn how to teach English to Thais. When qualified, he would sell his house and move to Thailand, maxing out his line of credit to raise cash before he left. There was no talking him out of it, as he had met the woman of his dreams, a twenty-one-year-old Thai girl who worked in a bar, and he was set on marrying her.

George, who was fifty-eight at the time, did exactly as he planned and departed the shores of home for Asia some months later. He walked into a good job at a Thai school the second day after he arrived (he still works there), and met up with his Thai beauty. It took him a year to adjust and realize the girl was just a barfly working in Bangkok. Today, eight years later, George has found a wonderful Thai wife who is also a teacher, and they live happily in a small house to the north of Bangkok. I see him from time to time, and have never seen a happier man and wife.

Summary

Finally, before you disappear, there are other questions you will seriously need to consider. In addition to your age and your physical condition, do you know where you are going and how you're going to get there? How are you going to raise the money? Will you need to change your identity? What happens if people come looking for you? What if you leave with nothing? Can you survive as a hobo? What plans do you have for your new life? All these questions will play a major part in where you go, how you get there, and what you do after you have disappeared.

With social networking, the Internet, credit cards, and so on, the world is not a wilderness anymore, and it's not that simple to disappear and never be found. For example, you would think it would be easy to disappear into vast wastelands of Australia, but it's not! Within a month of anyone going missing, more than 95 percent are actually located. Check out the link below, as this will give you some ideas: australianmissingpersonsregister.com

In order to disappear and never be found, it will take detailed planning and preparation—you will need a heart of cold steel to leave your loved ones—but most of all, you will need skills to make you plan work.

2

IDENTITY AND FALSE DOCUMENTS

Just before disappearing, it would be beneficial to create a new identity and name. If you do this discreetly, it will make it even harder for anyone to track you down, especially if you decide to stay in your country of origin. Remember, there are many factors that will aid anyone trying to locate you: your Social Security number, credit cards, driver's license, passport, marriage certificate, and the fact that you had an official name change can all be found, as they will be on record. With or without a name change, if you decide to remain in your country of origin, the chances of you being discovered are much greater. There are many ways to change your identity or create a new one: You can change your name, your appearance, remain in isolation, or go somewhere where no one will ever know your true identity. To start a new or second identity, it is possible in most countries to change your name, but all the rules that govern this vary dramatically. For example, in the United Kingdom, it is possible to change your name or any part of your name by deed poll, while in

America, the laws on changing your name vary from state to state and require some kind of public announcement.

If you intend to change your name, it is best to do this as discreetly as possible and not broadcast the fact. Many people have changed their name simply because they're in the theater, are a movie star, or are involved in the entertainment industry. Their pseudonym soon becomes accepted and is used to relate to that person. However, when you intend to use a different name in order disappear there are several factors that must be taken into account. If you are a normal person, then simply change your name without too many people knowing about it. On the other hand, if you are famous, just changing your name will not be enough as your face will also be well-known.

If you change your name to blend in with local people, many countries have mixed races such as Thailand and Malaysia.

Barry Davies

It is for each individual to decide if a name change is going to help once they have disappeared. In some cases, when you intend to disappear to a foreign land, it might be advisable to make your new name so that is in keeping with that country. For example, if you aim to travel to Malaysia and live in that country, you might call yourself Daniel Fuad or Ryan Chong. This way your new name will fit in with the community, just in case someone is looking for you by your old name.

Identity Theft

I shouldn't say this, but it's easier to steal someone's identity rather than to create a new one. While this is totally illegal, I write it here because it's an option that some people have used. Most people steal someone's identity to get access to their bank accounts, but what I am suggesting is stealing an identity to aid your disappearance. This means you are using their name to get false documents and travel under their identity.

An estimated nine million Americans were victims of identity theft last year, and the culprits were not your normal villains. The majority were opportunists who worked in mailrooms, hotels, and personnel offices—all places where you need to give specific information about yourself. Therein lies the key to all identity theft: information about the victim—it's everywhere: the social network, Internet, on bank statements, utility bills, etc., most of which ends up in the trash. A brief look at how many people live their lives is the best start to stealing someone's identity. At least 90 percent of people living in the Western world use a computer, laptop, iPad, or iPhone at work or in the home. On one of these devices will be enough information to help you copy the identity of someone.

If you choose someone of the same sex, physical build, looks, hair color, etc., it will make things even easier. All you have to do is step into his or her shoes. After all, you don't want to steal their money, just their identity.

How Do Thieves Steal Your Identity?

There are many ways to steal someone's identity: it could be by knowledge of the person, through work, a friend of a friend who talks too much, or a professional thief. Let's take a look at all the possibilities where information about you might be gleaned.

- Garbology (dumpster diving to you Americans).
- Personal information put up on social networks.
- Stealing your credit card details.
- Stealing your mail.
- By sending you a false email.
- Stealing your wallet.
- Going out with you on a date.

Usernames and Passwords

When using the Internet, no matter what you are trying to achieve, you will usually need a username and a password to log onto a site. True, some banks have a much more complicated system, but they are in the minority. Getting someone's username is fairly simple; it's the password that's takes a bit more effort.

Even so, getting someone's password is easy in most cases. I know that most of you by now have a whole list of passwords for a wide variety of Internet access points, such as your bank accounts, email, social networking, and so on. I would put money on it that you have a list of these passwords either on your computer or laptop, or that you use a similar password for everything? Am I right?

Now most modern day online facilities require a username and password, it's that simple; only a few of the larger banks have a more sophisticated login system. Barclays Bank in the UK, for example, has a series of checks that involve your credit card and a small handheld device, that together produces a different eight–digit entry code each time you log in. However, it is true to say that most logins are fairly standard, and once you know the target individual, the username and password are easy to guess.

The first thing you can do is to simply Google the target and get as much background intel as possible: parents, grandparents, children, hometown, pets, email address, and so on. Next, go to their address and steal the garbage. Quickly sift through the bin and take anything that looks relevant, such as scrap paper, utility bills, etc. Take them home and analyze their contents— you might need to do

Sorting through garbage reveals so much information about a person.

Barry Davies

this more than once. If you work in the same office or can gain access, check out their desk and waste bin.

When in position to observe a target's house to collect information, don't forget to look at their clothesline. Almost all households—including flats—hang out some washing during the week, and almost certainly on the same day every week. Observation of the clothesline over the period of a month can provide the following information: the number of people living at the address, their rough age, and their gender.

Try to gain access to the victim's email; you will need this when you want to get a new passport or driving license in their name. In order to gain access you will need to go online as if you were the target and ask to change the password.

The best way to learn about gaining access to other people's email and websites is to try it out on your own, i.e. do a self-check. To start, go to Facebook and try to reset your password. It's easy; all you need is your original password to get in, find where you change it, and then type in a new one twice. You can also access and change the email address and the name, etc.

Many log-in pages will also ask about personal details, such as your mother's maiden name or the name of your pet. If you have done your research on the target well enough, the answer to these questions can be provided in seconds. Even official information like a Social Security number can appear on many documents. A quick look at your personnel file at work will show you just how much information can be gleaned that could help anyone trying to steal your identity.

Receipts and invoices can provide so much information about your bank and your credit card. We all save our receipts for accounting purposes, or if not we put them in the trash, which means you are either hoarding information or disposing of it. Either way, it could be accessible to someone wanting you steal your identity. How many times have you been in a restaurant and seen people waiting

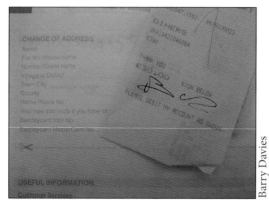

Barry Davies

Getting someone's signature and address is fairly simple using garbology.

to pay the bill, their credit card being waved around or placed on the table? How often does the waiter take your credit card away to scan it? Running your card takes time enough to snap a quick picture of both sides with the waiter's iPhone. Then, they not only have your card number and a copy of your signature, but also the three digit security number on the back.

Then there are the "skimming" devices, which will not only register your card details, but also the pin code when it's entered. Always be careful at ATMs, automated petrol services, etc., where there is a chance that a skimmer has been put in place. Check out the card slot with your hand, i.e. shake it and see if its moves or is loose. Check to see if there is a pinhole camera hidden at the top of the ATM catching your every move. Always cover the keyboard when you are entering your pin and always immediately report any suspicious withdrawal from your credit card.

> **Author's Note**: I have several bank accounts that have issued me credit cards. On each one, I insist they send me a text the moment I make a withdrawal.

Financial Resources

Financial resources will also play a big part in your plan to disappear. Not having enough money will limit your options, while having too much will bring you to prominence. If you are poor or of moderate means and have few assets that can be converted into cash, then your options to disappear are fairly limited. On the upside, very few people will miss your departure when compared to someone famous or extremely wealthy. Likewise, if you're poor and have not committed any major crime, there will be fewer resources spent on trying to locate you or to find the reason for your disappearance. This does not mean that you can't disappear; you could always go on the road and become a hobo or drifter, as this costs nothing.

If the police want you for a crime in your country or you are indicted on a crime and out on bail, the chances are that you will have surrendered your passport and your bank accounts will be frozen to stop you from disappearing. It makes no matter if you're innocent or guilty, the courts order these things done even in some divorce cases. So unless you saw this coming and made alternative arrangements, i.e. moved the bulk of you money into accessible cash and usefully applied for a second passport, your only option is to disappear as a hobo.

Good Financial Resources

If you are rich and intend to disappear to start a new life, then it will be considerably more difficult. You will have to be careful how you transfer your wealth so you can access it under a new identity. When you disappear there will be questions asked, and the amount of wealth you leave behind will determine the amount of trouble people will go through to establish if you are dead or alive. A wealthy person who has disappeared and leaves behind someone who will benefit from his or her wealth is asking for trouble. The beneficiary will do everything within their power to get you legally declared dead so they can get ahold of your money. That means you could lose all your money; on the other hand, if you transferred it overseas, it would merely indicate you intended to disappear. Transposing of your wealth will require some delicate planning.

If you are rich and want to disappear, the chances are that you have done something against the law! In such a case, you need to make sure you go to a country that is a tax haven, and one that does not have extradition treaties with your native country. This way you are fairly safe, but it can become expensive, especially in some of the more corrupt countries. You are best off buying yourself a new identity and foreign passport in a new name. Then travel as far away as you can.

Famous

If you are famous, your face will be public knowledge, especially if you have lived in the limelight of the media. If your face has been plastered all over the tabloids or on television for a number of years, then you will be easily recognized anywhere in the world. An

Amelia Earhart is a perfect example of someone famous disappearing.

example of this would be David Beckham, the British soccer star who has played for England, Spain, and America. His face is known worldwide. Without major surgery to alter facial features, it would be difficult for a famous person to completely disappear and never be found.

How a Change of Name Works in Law

In the United Kingdom, it is possible for anyone to change their name, although those under the age of sixteen will require their parents consent. In the United States, people can also easily change their name, but it should be noted that each state has a different law in regard to doing so. One of the best sites to download the forms is uslegalforms.com. Here, you will find the relevant forms for legally changing your name for most of the states in America.

Basically, it is possible for anyone to change his or her name, and in doing so obtain new documentation, such as a passport and driver's license, with a new name. By law, your name is legally established by usage, meaning that you have the right to call yourself whatever you want. Gradually, by usage and reputation, you become known by your new name. However, if you remain in your country of residence, many organizations will require you to produce some official documentation showing evidence of your new name and, moreover, that you plan to continue with it and not revert back to your old one.

You also need to know there are certain names that, while not breaking the law, will not be allowed. These include names that are:

- unpronounceable or incomprehensible
- extremely long
- containing numbers, e.g. Super8 or 4Real
- vulgar, offensive, or blasphemous
- chosen for purely commercial reasons
- chosen for a bet or frivolous purpose
- trademarked or subject to copyright, e.g. Coca–Cola or Asda
- a combination of names that make up a phrase or saying not normally considered to be a name, e.g. Happy Birthday or See You Later
- a presumed title—that is, a first name giving the impression that you have a title, such as Lord, Baron, or Princess
- giving the impression that you have honors, for example, surnames ending with OBE or VC
- a single name only—that is, a surname only, with no forenames

You might also find that some government departments have their own policy in changing of names, and this is also worth checking. To be on the safe side, it is always best to change your name to something normal. If you are changing your name and intend on remaining in the country, you will have to inform a lot of other people, such as the Inland Revenue or IRS, your doctor, and so on. However, as you are intending to disappear, it is best not to tell anyone else that you are applying for a passport in your new name.

Changing your name in America is just as simple as in the UK, but the laws do vary from state to state. I am sure I am correct in saying that in some circumstances, changing your name in some states can be kept secret if it is done to avoid further abuse by a pursuing partner. As in the UK, it is a felony to change your name to avoid debt, plus you cannot call yourself Prince Charles, Elvis Presley, or anything else to copy someone famous. Some people think that a stage name is real and legal: it is not. A person's name is the name given to them at birth until changed legally within the rules.

Finally, be sure to give some real thought to your new name. For example, if you intend to go and live in another country, such as South America or Malaysia, search for a list of common names from that country. Once you've made your disappearance and arrived in your new place of residence, that will make it harder for anyone trying to find you.

Driver's License

In America, driver's licenses are issued under different criteria depending upon the state you are living in; I would suggest you apply for an International Driver's License or an Inter-American Driving Permit. The process is very similar; you will need to supply two passport-sized pictures of yourself, a photocopy of the front and back of your driver's license, plus fill in the relevant forms and add the required fee. These additional licenses are good to have if you intend to disappear overseas or down to South America.

International Driving Permit details can be obtained from the following website: aaasouth.com/Travel/travel-drivers.aspx?nvbar= Travel:IntlDrivingPermit.

Inter-American Driving Permit can be found at: aaasouth.com/ documents/idp-app.pdf.

An International Driver's Licence is fairly easy to get.

Changing Your Appearance

I think most of us would agree that unless we had a face that is well-known worldwide, there is no reason for anyone to undergo plastic surgery to change their appearance. This drastic step would be painful and expensive. That said, there are numerous surgical procedures that can alter the basic way you look. For example, a facelift can make you look younger and even change the physical features of your face. While many people undergo plastic surgery to fix a physical defect or often to make them look younger, this kind of procedure is not recommended if you simply want to disappear.

However, should you want to go down this road, one of the best hospitals for plastic surgery at a reasonable price can be found in Bangkok, Thailand and is called Bumrungrad. This hospital offers a wide range of surgical procedures including plastic surgery and reconstruction. The standards of this hospital are really world–class and on a par with any hospital in the West. In addition, they are quick and discreet. Visit http://www.bumrungrad.com/en/plastic-surgery-thailand-bangkok for more information.

Remember, if you change your facial appearance, you should also change your main means of identification, especially your passport. For example, if you grow a beard and a beard does not show on your passport photograph, you will not be allowed to enter the country unless you either remove the beard or get an updated photograph. This also applies for any major facial changes, piercings, or tattoos.

Disguising your face is quite easy if you wish not to be discovered in public: you can wear a wig, put on a pair of glasses, or use makeup to alter your face. There are benefits to carrying some extra clothing to disguise yourself too, especially when you think you're being followed.

False Documents

Disclaimer: The information in this book relating to gaining a new passport by deception or by stealing is purely for information. I do not advocate stealing or obtaining a passport or any other legal document, and mention it here purely in context with the what–if scenario within the parameters of this book. All of the information below can be freely obtained from the public domain.

There are lots of ways to get false documents, but I have to warn you that it's getting more difficult to fool the system. In this modern age, it is extremely difficult to make a false document that will pass the sophisticated scanning devices used by the authorities, especially when it comes to identification documentation. Also, if your document does show up as false, you are in for a lot of trouble— particularly at airports, where you can get both fingerprinted and have an iris scan. Spies and agents engaged in subversive operations are often obliged to use false documents, including passports and identity papers, but these are normally produced by government agencies by simply modifying a real passport under a different name.

False documents that are not used as a means of identification are easier to produce, and the chances of being found out less likely; for example, driver's licenses or birth and marriage certificates. For what it's worth, my advice would be to get hold of a genuine document and either convert it or pretend to be that person.

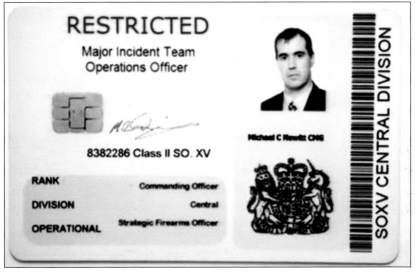

The ID card used by Michael Newitt that even fooled the police.

With good documentation and a strong cover story, it is possible to get away with just about anything. An English man named Michael Newitt lived a fantasy life; he called himself Commander Newitt and posed as a senior police officer. He pretended he was an undercover officer mainly to stave off companies to which he owed money. His identity was backed up by a series of impressive items, such as an ID card and police wallet. He not only fooled civilians, but also the police for several years.

Counterfeit money is also produced in large quantities and used by many government agents or counterfeit organizations, often as a means of subversion or to weaken an economy. Most intelligence agencies refer to this type of forgery as "repro," with the person who makes the forgeries known as an "artist." Most agencies have their own artist department responsible for acquiring documents and tailoring them to fit the agent for any special operation. Of these, by far the most important document is the passport. However, unless you work for a government agency or know some very smart people, I would not even bother to try and forge a passport in this digital age. Even the experts sometimes get caught trying to enter a country on a forged or false passport.

In March of 2004, two Mossad agents, Uriel Kelman, 30, and Eli Cara, 50, were jailed for six months in New Zealand for trying to obtain false passports. The plot was discovered when a passport officer noticed that a passport applicant was speaking with a Canadian or American accent. The clue led to an investigation that uncovered a complex conspiracy involving up to four Israeli agents. Using a fraudulent birth certificate, they were attempting to create a false identity for thirty-six-year-old Zev Barkan, another suspected Israeli spy. Officers planned to arrest the spies as they picked up the completed passport. However, Cara had preempted this by having it sent by courier to an apartment block, where it was to be collected by a taxi driver and taken to a rendezvous with Kelman. Police surveillance caught Cara acting suspiciously, close to the central Auckland apartment block while Kelman was arrested after fleeing the other rendezvous and throwing his mobile phone into a hedge. Both were sentenced to six months in prison for their involvement in the plot.

A New Passport

The secret to making a new identity is to steal one from someone else or someone who is dead; the more recent the death, the better your chances of accomplishing this. You will need to scour the obituaries and look for someone of the same race, age, and gender as yourself. It is best to look in a large city where the death rate is higher than in a country village. Once you have located a match, you will need to gather information about the deceased and obtain a photograph if possible. Normally, the deceased's address can be gleaned from the newspaper, and if the death is very recent, there is nothing to stop you from going along to the house and pretending to be an old friend. Once there, you can simply ask for a recent photograph as a keepsake. If you discover that the deceased lived alone, you might try a little burglary. If you are lucky, you might even turn up a passport or birth certificate (relatives normally dig these out when someone has died). Be careful not to steal anything else.

Another scam used by people trying to obtain a new identity is to pretend to be a representative of the local coroner's office (have a fake ID), and make an appointment to visit the bereaved family. They usually make sure their telephone call is just after closing time for the coroner's office, so there can be no back check by the family. They simply state that they will need to see some relevant documents the family can easily find, such as the Social Security number, passport, birth certificate, etc. As most people rarely deal with the coroner's office, they will think this is a normal procedure.

If neither a burglary nor a scam manages to produce a birth certificate or passport, then you will have to think about making and obtaining false documents. This is not as big a problem as it might seem; once you have established the deceased's details it is a fairly simple matter, but one with risks. You will need to obtain a legitimate copy of the deceased's birth certificate, and then apply for a new passport—this time using your own photographs.

In many cases, the deceased may look nothing like you, but this is not a real problem, as others have shown in the past. Most people have used a computer morphing program to scan a facial picture of them and that of the deceased, then the program merges them halfway. Many examples of this practice can be found on the Internet. The final result should look something like you, but it could also be the deceased. They then print out four good passport photographs using photographic quality paper from any computer shop and include these with their new passport application.

When in possession of the deceased's passport, some people have changed their identity by requesting a second one from the passport office. While many countries will issue a second for business purposes, the applicant will need to prove this in the form of a supporting letter from his or her company or organization. In order to age a new passport, people have used lots of fake travel visas. They easily copy foreign visa stamps from one passport to another. They do this by using a flat tray filled with a half inch of set gelatin. They then place a page of the well-used passport onto the gelatin, pressing it down hard, which will leave an impression. This impression is transferred to the new passport pages. They simply repeat the process using a clear area of gelatin each time to copy and transfer old visas.

They will end up taking over the identity of the deceased person who, although dead, did exist. Providing they use their new identity in a foreign country, the chances of discovery are minimal. They will be able to open bank accounts, buy a home, and do very much as they like.

There are many other ways of obtaining a false passport; stealing one while abroad is the easiest, although this will be reported and canceled. Another way is to purchase a flight ticket under a false name using some form of ID (but not a passport); this is best done through a travel agent. Some people claim to have had their passport and wallet stolen and go to the local Embassy with four photographs of themselves to ask for a new one. They state that they are leaving the country the next day, explaining their flight ticket is all the proof they have. With a little luck most will get a new passport without too much hassle. However, in some cases, especially the European Union, you may only be issued with an Embassy paper that will allow you exit into your own country.

Although any passport will do, it is generally best to get one from your own country or a close neighbor that speaks the same language (an Irish man using a British passport would be a good example). This will forestall any problems at the embassy if they ask you any questions.

Getting a Passport Illegally

If you wanted to get a spare passport with someone else's name but your picture on it, you would either have to buy a passport off some shady character or steal one. People lose or have their passports stolen every day; while some are recovered quickly, others find their way into the hands of criminals. It could be long and drawn-out pro-cess of getting ahold of a stolen passport this way, and it could cost you a lot of money and lead to other problems that I will not go into. Most people who are desperate for a passport generally steal one or, better still, have someone else steal it for them. This is a little like identity theft mentioned earlier, but a much easier.

Having your passport stolen causes a lot of problems. Make sure its kept in a safe place when not in use.

People have deliberately stolen passports in any number of cir-cumstances. Airports are very popular, but there are many other venues such as any holiday destination where young men and women let their hair down (get drunk) and have fun. For example, in Greece or on the island of Ibiza, it would be easy to target a male or female and steal their passport, which is a daily occurrence. These holiday destinations are renowned for wanton revelry where being drunk and picking up strangers to have sex is the norm.

Would-be passport thieves simply pick a target; someone who looks like the right age, build, looks, sex, and nationality. They venture out around 2:00 a.m. to see which of the revelers are hardened drinkers, then watch and follow them as they stagger home to their apartments or hotel rooms. In Thailand, it would be a simple matter to employ a street girl to do the job. These are girls who are at the bottom of the sex industry—they do not work in bars but ploy their wares openly on the street. Many of the older, wiser women will wait for a drunken guy to come along and pick them up; they then take the man to his hotel or room. While he is asleep they rob him of everything they can.

Author's Note: I know a British man who has worked in Bangkok for many years. Each Saturday will find him drinking in the bars up Soi 5 of Sukhumvit in central Bangkok. While he starts the day off with his friends, toward late evening he is alone and very drunk. On many occasions he has been picked up by street girls who have taken him back to his apartment. Several times he has woken in the morning to find half his apartment missing, including the furniture. To my knowledge, he has applied for at least four new passports.

The passport thief would then point out his or her target to the woman and get her to pick them up once they have had enough to drink. Of course, the thief would have to pay the woman something in advance—she works better that way. They would specifically ask her to get the man's passport and any other documents, such as a driver's license. With the promise of a lot more money if she is successful, you can bet she will fulfill her mission.

Once you have a passport, use some sandpaper to disfigure the picture; enough so that water can penetrate the actual photo in the passport. Then put it in the pocket of an old pair of trousers and wash them in a machine. Check the state of the passport; the aim is to keep as much of the writing legible while destroying the picture. Once this is done, simply take the passport to the local Embassy and ask for a new one, explaining you accidentally put it in the washing machine. If the passport was stolen in Thailand, then take a quick flight down to Kuala Lumpur in Malaysia and request a new passport there. This is because at some stage the person you stole it from will be reporting this to the Embassy in Bangkok, which could start alarm bells ringing.

An alternative would be to find a look–alike and use his or her passport without changing anything. In essence, you would become that person by using his or her name. Again, some background information on the passport holder is always very useful.

ID Cards

Identity cards—such as a driver's license—do not pose a serious problem as they are much easier to forge. There are many Internet sites offering a whole range of identity cards; however, few of these are any good. Many people prefer to make their own, and all they require is a modern computer set up and a scanner, which will help them make a fake copy of just about any paper object.

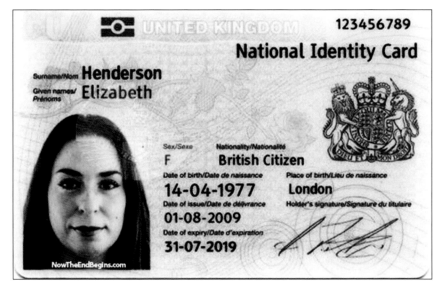

This is a fake ID as the United Kingdom does not have a national ID system (yet).

They simply scan a similar legal document using a high resolution scanner (at least 600 DPI, or dots per inch) and always use color. This will create a large file, so they will need a good computer and graphics package to handle it. Once they have a copy, they simply manipulate it to suit their needs. This might mean removing another person's photograph, name and address, and replacing it with their own. Most good graphics software will allow them to do this with ease.

Next they will choose the correct type of paper to print on. ID cards are normally printed on thick card, whereas birth certificates are on fine paper. Ideally, the paper should be the correct weight and color, with no watermarks and as close as possible to the original. They then simply print out their fake, back and front if required.

If their fake copy needs a watermark, they will need to prepare the paper by embedding a watermark picture to the same density as that of the original. Most modern graphics programs allow them to do this. Simply chose the correct design, select the density and size, and print. They then use the watermarked paper to print their fake copies.

The problem now is to process their new fake so that it looks original. If it is an ID card it will need trimming with a scalpel and sealing in a plastic protective jacket. Most forgers will use machines that can be found in most major office supply stores. A simple visual check of the original against the fake is all that's needed to complete the forgery.

* * *

Older documents are aged by putting a damp (not wet) cloth over the paper and placing a household iron over it several times. They then place the document in the sun for several days until it has faded to their satisfaction; folding it several times before dusting the document with a little cigarette ash over the surface to add that final authentic touch.

If you think the above method will not work, I can sure you it does (most times). I had a dispute over land I own in Spain, and before I went to court, I had to prove something that had happened five years previously. My Spanish solicitor typed out the document, back-dated it, and had me sign it. He then crumpled it up and put it outside in the hot sun for several days, resulting in a document that he could use in court to prove my case (I was legally in the right, but this practice, I was assured, is normal).

Looking Like Someone Else

If you have resorted to stealing an identity, you might as well do it properly. The key thing when stealing someone else's document is to make sure they have very similar facial features. If you can get a fairly close match, then you don't need to change or adjust anything; just be sure that the person's name appears on either the passport or driver's license.

> **Warning:** Do not try to go through an automated passport control where the machine takes your picture and matches it to the one on your passport, as this will raise an alarm. It is best go through the manned system. Providing your face is a close enough resemblance, you should have no problems.

It is because pieces of paper or plastic are relatively easy to counterfeit that governments have introduced more secure methods. Not all countries have an e-passport or "chipped" passport system, and many still have the older bar–coded method. However, all the chip actually does is verify that the picture and information on the ID page is the same data that is stored on the chip. Some people refer to this as a biometric passport, but they are wrong—a biometric passport can also hold things such as your fingerprints.

3

MONEY

A banker once told me that during any crisis, economic depression, or if you are trying to evade the law or a chasing wife, "cash is king." This is certainly true if you are wanting to disappear, as you will not get very far if you don't have some money; even posing as a hobo will require some finance. So from where will you get your money?

Most people will have a little money put aside, whereas others may have a lot. This next part has been written for those who have very little. Your plan for disappearing should have a section on harvesting funds, and you should give yourself at least a year in which to do this. There are a few things about cash that you need to be aware of: loose money can get stolen, lost, or destroyed by fire, flood, etc., and once it's gone, it's gone. However, if you play it safe and keep your money in a bank it will be secure, but it will also be traceable. My suggestion, therefore, is to use a safe deposit box.

To begin, if you have savings—either joint or personal—slowly siphon them off over the period of planning your disappearance. There are many ways of doing this: one that we have in the UK is called "cash back." While I am not sure how it works in America, in Europe, cash back is available in supermarkets. It simply means that when you pay

Cash is always king, but make sure you have it in recognisable currency such as US dollars, British pounds, sterling, or Euros.

for your goods at the checkout, the assistant will ask you if you would like cash back. The money you ask for is added to your food bill. It's just a convenient way of saving time in front of a busy ATM machine. It's also a safe way of siphoning off funds from your bank account without anyone paying attention. True, the amount of cash back is limited to £50 per visit, but there is nothing to stop you from visiting several supermarkets a number of times in the same week. Over the course of a year, it will all add up. If people do get access to your bank account details after you have disappeared, then they will not be any the wiser as a credit card transaction in a supermarket will go unnoticed.

Other ways of raising cash can also be exploited, such as earning extra money by doing small simple jobs that pay just a few dollars at a time. Buying and selling surplus items at a car boot sale or on eBay will do very much the same. Here are a few examples of how to harvest extra cash:

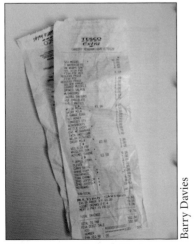

Barry Davies

Most supermarkets will give you up to £50 ($75) as cash back when you go shopping.

- Get a second job. This might be something like a simple cleaning job once a week for a few hours, but if you save the money over a year it will help.
- Weekend markets, garage sales, car boot sales—sell off anything you will not be taking with you once you disappear.

- Advertise larger or more expensive items on eBay.
- Cut grass or clear snow.
- Teach what you know, i.e. language—target foreigners entering your country.
- Do online consultancy if you're very knowledgeable on any subject.
- Edit books for publishers, or proofread for authors.
- Babysit goldfish, cats, dogs, and any small animals for people on vacation or who are infirm.
- Offer airport runs—be cheaper than the local cab.

Never use your disappearance cash fund for an investment to add interest, for as soon as you do this, it can come to the attention of the tax man. Always keep in mind that you do not want to leave any trace behind indicating an intention to disappear. Make sure that, when you have fallen off the earth, there is no trail to follow—leave everything in order as it would have been if you had not intended to disappear. Continue to pay your credit cards on time, pay any ex-wife whatever she is due—do nothing out of the ordinary. Harvest all the small amounts of cash and keep them in a safe place so that when it becomes large enough you can put it in a safe deposit box. True, you will be missing out on interest, but your money will be secure.

Once you have acquired a substantial amount, you might try to legitimately go on vacation and take the bulk of the money with you. You may tell family and friends that you want to visit your long lost relatives in Ireland, Germany, or the UK; you don't really have to go to any of these countries, just say that you are. Who is going to check? Once you get on the aircraft, you can go anywhere. Most European countries will happily open a bank account for you, or you can purchase a pay-as-you-go credit card. In fact, you could buy several pay-as-you-go credit cards then simply go into any post office and top them up evenly with the cash you brought with you. When you return home, try a simple withdrawal on each card to make sure they are working. You can monitor your accounts online. Now you have some safe money in a foreign bank, which no one knows about, and if you choose the right prepaid credit card, you can withdraw local currency in just about any place on the planet.

While it is possible to pay people in cash, when you need to purchase something via the Internet, you will need a credit card. Again, the best kind of credit card is a prepaid one, which is untraceable.

Credit cards are a virtual necessity in today's technology–heavy world.

Credit Cards

There are a couple of things you should know about normal credit cards. One is that, while they can be traced through transactions, the only people who can gain access are either the original creditor (i.e. the credit card issuing bank) or the government if you are of interest to them.

Prior to your departure, you should use your credit cards as normal to avoid any suspicion. If you have a large sum of money in the bank, you can always use your credit card to purchase items that you can then sell for cash. While you might lose a little on this type of transaction, you will be converting your savings into your untraceable fund.

Most people think that cash is untraceable, but they would be wrong. Many currencies now have RFID chips in then, the euro being just one example. If you don't believe me, try putting a 5 euro note (smallest denomination) in the microwave for a few seconds and see what happens. If you check the foil strips, you will see a small burn mark where the RFID chip has fried.

Prepaid Credit Cards

There are many ways to buy prepaid credit cards, although the law varies from country to country. Also, there are no restrictions on you buying a prepaid credit card in one country and using it in another. You simply have to make provisions to top the card up at a convenient place using cash. Prepaid cards are typically marketed to those who have no credit history or have bad credit; they are also used for young people who don't qualify for standard credit cards. You should take care with which prepaid card you select, as many are loaded with excess fees. Prepaid cards are not covered by the card act that regulates credit cards, therefore many of the guarantees and protection that come with a standard credit card are denied the user of a prepaid card.

In the UK, one of the best prepaid credit cards is from Virgin. A Virgin Prepaid Card costs just £9.95 to buy. You have a choice of two tariffs, one of which being pay as you go. If you think you will only use your Virgin Prepaid Card occasionally, this tariff could be best as there is no monthly fee. You just pay 2.95 percent for each card transaction and cash machine withdrawal in the UK. The other option is a pay monthly card. If you are planning to use your Virgin Prepaid Card a fair bit, this tariff might be more suitable as there is no transaction fee for purchases in the UK. There is a flat monthly fee of £4.75 and a fee of £1.50 for each cash withdrawal in the UK. Loading and reloading your Virgin Prepaid Card with money is easy—it works just like a phone card top up. There are lots of ways to top up, and more than 34,000 places in the UK to choose from, including any post office, by debit or credit card, direct transfer from a bank, etc.

American Prepaid Credit Cards

So why go to all the trouble of gaining a prepaid credit card overseas? The answer is simple: To help the government fight the funding of terrorism and money laundering activities, American federal law requires all financial institutions to obtain, verify, and record information that identifies each person who opens a card account. What this means for you is that when you open an account, you will asked for your name, address, date of birth, and other information that will allow the agencies to identify you. They may also ask to see a copy of your driver's license or other identifying documents.

Additionally, in America, the maximum amount that can be spent on your card per day is $2,500. The maximum value of your card is restricted to $9,999. Many foreign counties issue prepaid cards with the only limit being the amount you have in your account.

Setting up an Offshore Bank Account

An offshore bank account is simply using a bank in another country. Once you have done this, you will have access to funds anywhere in the world, providing the offshore account issues you with a credit card for the account.

You can set up an offshore account without actually leaving the country by using the Internet, but my personal advice would be to visit the country where you intend to locate and set up the bank account personally. You can use the travel section of your disappearance plan to organize one trip overseas to set up and test your offshore account. In reality, it makes no difference where you set up your account, but it is advisable to have it in the same country you intend to live in once you disappear or very close. A good example would be to set up an account in Malaysia while actually disappearing in Thailand. This would give you direct access to your account and you could visit the bank from time to time should the need arise.

Naturally, you should select a country that is politically and militarily secure and where the banking rules are fairly simple. Once again, you will need an address, but it is easy to setup a postal address and have your mail forwarded to anywhere in the world. Better yet, opt to run your account via the Internet. You will also need proof of your identity, and again, this is simpler if you present yourself in person. They will simply take a copy of your passport and any other documents required; these will sit in a file that no one will ever search. You may need a valid reason to open an account in some countries, but you can always say you are representing your company in setting up a new regional office or retiring to the country. The process should be fairly simple, and all you need to do is to make a deposit—cash is best—and request a credit card.

Once your account is up and running, put in as much money as you deem necessary. My advice is to open at least three accounts—either in the same country or in neighboring countries—and spread the risk in case of something going wrong or an account being suspended for any reason.

Being British, I am not sure how an American citizen stands in regard to foreign banking, but my understanding is that you are NOT allowed to open a foreign bank account without declaring it (so much for the "land of the free"). When it comes to purchasing a prepaid credit card, there is absolutely nothing to stop you, although in effect you are actually putting your money in a foreign bank and simply withdrawing it from your card anywhere in the world.

Offshore banks do not report your income to the IRS like US banks do, so they use an honor system, which means that it is up to you to report your income. It is possible to open an account in Switzerland via email, but turning up in person is a much better idea. Bear in mind, though, that in recent years the US has made inroads with the Swiss banks and could possibly trace your account. They are not as private as they used to be. My understanding is that one of the best banks is Credit Suisse, which will open an account for nonresidents. There is no minimum deposit, but you will have to pay annual fees unless you have more than 50,000 Swiss francs in your account.

A few years ago, you could open a bank account in Thailand with ease. However, in the past four years things have changed, and you will now need a work permit to open one. On the other hand, opening a bank account in Malaysia is a straightforward process. Bank accounts are not confined to Malaysian residents, and foreigners can freely set up bank accounts in Malaysia if they fulfill certain requirements.

Indian banks are fairly flexible and it's easy to set up an account.

In order to open a bank account in Malaysia without holding a residency permit, you will need reference from either a Malaysian contact or your company. It is nominally required for you to hold a work permit or to be in Malaysia under the MM2H program to be able to open a bank account. But this does not always matter because, as regulations are not that firmly set, some banks will still accept your application without such a visa. Take your passport and a simple letter from your employer (it's simple to get one in

Malaysia), and the rest is easy. However, once you have opened your bank account, getting a credit card in Malaysia is not as straightforward. You will need to question a prospective bank about it before you finally settle on one. These restrictions aside, you can still buy a pay-as-you-go credit card from several banks in Malaysia. The following web address will take you directly to one such facility: http://www.ambankgroup.com/en/Personal/Cards/PrepaidCards/Pages/default.aspx.

Before you finally disappear, you need to convert as many of your assets into cash as possible, but do it in such a way that will not arouse suspicion or leave a trail of evidence to indicate your intention. There are many ways to gather cash, and the amount you will be able to gather will depend on the assets you have.

If you have cash in a savings or deposit account, you should slowly remove this by making small withdrawals over a period of time and place the cash in a safety deposit box. If you do this cautiously, as if you are using the money for day-to-day expenditure, then there is nothing to indicate that you withdrew any large amount in order to disappear. If your employment involves a lot of overseas travel, you can withdraw a decent amount of foreign currency each time, and then when you return, simply convert your unspent foreign currency back into cash. Your expenses will show you traveled overseas and will therefore legitimately hide the transaction.

Don't forget, before you disappear, you will want to sell as many of your personal items as possible. If you intend to do this, make sure you choose a weekend market that is some distance away from your home, so the likelihood of anyone recognizing you is limited. An even better way to dispose of goods is to sell them privately on eBay. The key to remember is not to sell or do anything financially in a way that will cause people to wonder what you're up to. If they do ask a question, simply say something like you have sold your motorbike because you are thinking of buying a new one.

Car boot sales are always a good way of converting your old belongings into cash.

Investing in Gold

Gold has been used as currency by most of the known world and still is today. Gold is a good investment, acceptable just about anywhere, and has a great liquidity (it's easy to sell). You may consider taking up a hobby collecting gold coins. You simply keep your new activity a secret and, if discovered, simply say your secrecy was to protect your investment from robbery. The American Eagle is one such coin, as it shows its value and can be exchanged for cash with confidence. You can purchase American Eagle Gold Bullion Coins from most major coin and precious metals dealers or via the bank. Many dealers advertise their coins online, and a simple transaction can secure your purchase.

Gold coins are accepted by most people around the world—some people prefer them to paper money.

Gold coins are also easy to transport between countries, as they are small and not too heavy to carry on your person if you only transport a few at a time. Additionally, most banks will accept recognizable gold coins as a deposit. On operations in foreign countries where the British SAS are required to work behind the lines, they are issued with a money belt that contains twenty gold half sovereigns with which they can buy their freedom if captured or evading the enemy.

Transporting Your Money

If you wish to carry large sums of money on your person when you are traveling overseas, be aware that most countries operate strict rules on the amount of money you can actually carry in and out of the country. Usually these questions are asked on the immigration and customs form you're required to fill in prior to entry. However, if you behave in a rational manner and provide no reason for the customs or immigration officers to search you, then there is a good possibility that you can carry a large sum of cash distributed throughout your personal clothing and your luggage. When doing this, do not try to hide the money in a covert way because if customs searches your luggage and finds any money hidden in an unnatural way, they will strip search your entire body. My advice is to carry as much you can in high denomination notes in your wallet with an equal amount in a plain brown envelope. This way you will avoid any suspicion, and if the envelope is discovered you simply say that this is a deposit for something you wish to buy in the country.

Author's Note: Legally, when you travel from country to country you're required to declare cash or other monetary means such as traveler's checks. However, you do not have to declare gold. This allows for another possibility whereby you purchase gold coins in your own country and sell them in another. A stopover in countries such as Saudi Arabia, Oman, or Kuwait will allow you to buy gold at a good rate with US dollars. You can then take them on to Singapore or Malaysia and sell them with little loss (and possibly some gain) in the exchange there.

Hiding Your Cash

Offshore accounts can possibly be traced, and one can lead to another. Personally, I think having cash in US dollars or Euros is always a good thing. Once you have disappeared, make sure you always have enough cash to leave a country in a hurry, either by plane, ship, or overland. It is risky carrying large amounts of cash around with you, so don't keep it in your wallet; either wear a hidden money belt or leave it in a secure place that you can access quickly.

Hiding your cash is easy if you use a little imagination.

When it comes to actually transporting your money, you will want to keep it safe. Traveling with money is and always will be a hazardous business, but if you take care and follow a few basic rules, you should be safe enough. The first rule is NEVER ever let anyone know you are carrying large sums of money. The second rule is to ALWAYS carry a second dummy wallet in case you are mugged.

Dummy Wallet

If you should find yourself in an unsavory district where the rate of mugging is high, you should always carry a spare wallet. Fill this with a few dollars and some out-of-date credit cards. If you do get mugged, remember the mugger will want your wallet in a hurry and then run off. Always hesitate a little if mugged, don't be too hasty in offering them your wallet; let them demand it first. When you do give it to them, throw it on the ground and back away. They should run off without examining the contents—make haste and go the opposite way, just in case they come back.

Never carry more cash than you actually need. Withdraw only what you need from an ATM, and don't let anyone see how much you have taken out. At the first oppor-tunity, split your cash into several small piles and secrete them about your person or stash them in a safe place.

Here are a few ways to hide your money; they are not fool-proof—a professional or customs officer would find them quite quickly. Remember, when they find one hidden stash, they will rip you apart for the rest.

Make up a dummy wallet to prevent your real one from being taken in a robbery. This one contains only useless cards and old foreign currency.

- Utilize your pants. Put your money or credit cards in a small snap–sealed plastic bag down the front of your pants. Customs officers will happily feel your bottom but rarely put their hands on your crotch. Ladies can also try under their breasts; again few customs officers will go there unless it's a major search.
- Wrap the same plastic bag around your ankle and secure it with an elastic band.
- The waistband of a man's trousers is one of the best places. Don't overdo it though, and spread the money out evenly along the whole waistband.
- Make a hidden pocket inside trousers, shirts, or jacket as a last resort.
- Use a large non-padded dressing that will cover your money and stick it to your body. Dress and test to see that it's undetectable.
- On the soles of your feet, but make sure you cover it by wearing socks in case they make you remove your shoes.

We are all used to being searched today due mainly to the ter-rorist threat at airports, but even the most vigilant of airport secu-rity has its limits. We have to remove our laptops, jackets, belts, and

Airport security is very strict. Denude yourself of any metal, and you should pass through with ease.

shoes. You then enter a machine that will normally only check you for metal, but recently these scanners have become more sophisticated and are able to detect drugs and explosives. Luckily, we are not at the stage where they can detect money.

You can help yourself by not wearing any metal—no watch, no jewelry etc.—and there is a very good chance that you will go through the scanner with little or no trouble and without the alarm being set off (normally these are set to high anyway). If you do get patted down, they always miss the following places: your crotch, female breasts, head, mouth, anus, the very bottoms of your trousers or jeans, and soles of your feet.

Summary

As different rules and laws apply to each country, it is difficult to specify exactly how you should first harvest your disappearance fund and how you should then conceal it from others. The odd thing about money is that it does attract a lot of attention, especially when a country's tax or revenue department wants its share of your earnings. Simply attain as much cash as possible, either from your own savings or investment or by earning extra money.

Once you have what you consider a reasonable amount, make sure you convert it into some secure but readily accessible source, such as prepaid credit cards or gold coins. Be very careful not to have all your eggs in one basket and to separate your disappearance fund into several different accounts. If you're transporting money, it is best done by having it on your person or hidden in your on-board luggage—always keep it within sight.

Your disappearance fund is vital to your plans. Even if you choose to become a hobo for several years, you will still need some financial support.

4

RESEARCH AND PLANNING

There are two ways of looking at your disappearance plan:

1. You make preparations to become someone else.
2. You remain your original self and simply disappear one day without reason.

There are good and bad points to both. Should you wish to change your name and obtain a new passport, remove as much information about yourself from the Internet as possible and make sure you leave no clues that you planned to disappear. If, on the other hand, you do nothing—continuing your life as normal until the appointed day and then simply disappearing—you will leave no clues as to why you went missing. In either case, you will need to do some research and plan how to best succeed with your disappearance.

Planning your disappearance so that you will never be found is a difficult task; time is needed to get everything in place: finance,

Planning is like a puzzle; you have to get all the pieces to fit properly to make it all work.

cover story, and your new beginning. How and when you decide to go will very much depend on your own personal circumstances. For example, if you are in a really bad relationship where physical abuse is a constant occurrence, you may need to disappear sooner rather than later. If your life is simply going nowhere and you seek a fresh start, then you have time to plan in more detail.

Planning

Planning basically means gathering as much information as possible to aid in your disappearance, which will hopefully allow you to stay below the radar and never be found. You acquire this information through research. You may wish to know how to change your name, how to apply for a new passport, or what the possibilities are for finding legal work in a specific South American country. Your outline plan will help you build a list of questions you must ask yourself. These might include:

- Why are you going?
  You should ask yourself this question very seriously—what is the problem the drives you to want to disappear? First of all, you should confront yourself and your fears and decide if disappearing is the right thing to do.
- Who do you leave behind?
  There is always someone who loves you—mother, father, brothers and sisters, friends—and you must seriously think of the effects you're leaving will have on them. In many cases, some of these people will wholly rely on you for support.
- What do you leave behind?

You may leave nothing behind, but even those with wealth can lose everything once they have disappeared.

- Can you clear your personal history as far as possible?
  Everyone has a personal history from the moment they are born until the day they die; in this digital age there is more personal history publicly available than ever before. Removing this is not an easy task.

- Are you capable of using deception and disorientation?
  Of all the parts of your plan, deception and disorientation are by far the most important. The deception will stop people from looking for and finding you.

- Where will you go and why?
  There are many choices for you to make when deciding where you will go. You may plan to stay in your own country and simply change your name or the way you look, or you may go off to a foreign country and seek a new beginning altogether.

- How will you get there?
  The modern world is a transient place, and it's possible to travel to almost anywhere on the planet. However, many of these methods of travel require documentation that can be verified and traced.

- What finances and resources will you have in place when you leave?
  It is not easy to disappear without some financial resources, especially if you plan to travel to new country. But with-drawing large sums of cash, selling off assets, and harvesting money may all help indicate that you intended to disappear in the first place.

- What will you do when you reach your country of destination?
  In many cases, people will go to a new country, but you must plan what you will do once you reach your destination. For example, can you speak the local language, is it possible for you to get work, and can you comfortably start a new life in your chosen country?

- How will you avoid the past catching up with you?
  In the case where your disappearance may lead the authorities or investigators to locate you, you will need to delete any traces that they could possibly follow.
- What skills you will need to acquire?
  Your disappearance plan may require you to learn a new set of skills that will enable you to survive. For example, if you intend to live rough, you need to know how to become a scavenger and live off the land. If your plan is to go and sail the oceans, then you will need some skills in sailing and navigation.
- When to leave?
  The best way is to simply leave at the most innocent moment. For example, in a restaurant with friends, excuse yourself and go to the toilets, never to be seen again.

As an example, this is what I would do if I were to make a plan to disappear and never be found. First and foremost, let's say I have an overwhelming desire to disappear and forget about my family and friends, as this is one aspect we each have to come to terms with individually.

When to Disappear?

This must be a primary factor in your disappearance and there are several ways of doing it. Whichever way you choose, you must make it look real, provide a genuine motive if planned, or simply just not be there one day. If you only have the briefest of time to plan your departure then it has to be good. Below are a few examples of when to disappear.

- Scenario 1:
  You are a downtrodden, abused forty-year-old housewife and need to get out of your situation. You have $750 in your joint account, $500 of which you put on a prepaid credit card yesterday. Without warning, you go to the supermarket on

Finding work overseas, where you only return home every few months, is a great way to slowly disappear, as those back home start to live their lives without you.

the pretence of doing the week's regular food shopping. That morning you pack whatever you need into a small suitcase without anyone noticing. Your husband goes to work or lies on the sofa watching television. The kids have grown up or gone to school. You drive to the supermarket and withdraw the remaining $500, fill the car with petrol, drive out of state, and never look back.

This requires little or no planning, but the chances of success of you never being found are slim. You could drive more than 1,000 miles, then dump the car at a small motel where you check in. Your money supply is limited so you will need to find work. Whether you are found or not will depend on if your husband and family want to find you. The only upside is you will have some time to yourself and be away from the abuse.

- Scenario 2:
  You are a single thirty-year-old man; you work hard and are a good guy, but your wife has left you for another man

and taken the children with her. She is very beautiful and thinks she can do better than you as a husband, so she files for divorce. You decide to disappear prior to the settlement, having made a detailed plan that involves you hitchhiking across Australia. You harvest as much wealth as possible, including the family home, car, etc. Prior to the divorce hearing, you fly to Australia to begin your two month walk—no one hears from you again.

Literally thousands of people disappear into the Australian bush while trekking every year. Most of them are found, but only because they want to be.

This plan requires a lot of careful planning and near perfect timing. The first thing on the list is that you secretly set up home in the next state, change your name, and only inform those who really have to know. With a new passport and driver's license, together with your accumulated wealth, you head for Darwin in Australia to start your trek. But you don't go trekking. Instead, you head for the harbor and look at the bulletin board at the local marina. There you will find vacancies for boat crews: some are "work your passage" and others require a small fee for your food. You get yourself to Indonesia and make your way to Jakarta, being sure that when you land you get some form of immigration stamp (otherwise you will have trouble leaving the country). From here on your trail will have gone dead—keep it that way.

So it is possible to just walk out of a situation; equally you can plan your disappearance in detail. If you can, start your plan to disappear at least one year in advance, eighteen months if it's feasible. Next, play down or pretend to repair any current problems you might have, be they financial, job related, domestic, or romantic—allay any suspicion that you might have cause to disappear. Next on your list would be to consider where you would like to go and how you are going to get there. Your destination will impact the resources you will need to take with you. For example, could you find work in the country of your choice? How much money will you need to tide you over until you are settled? Are you capable, and would you be happy, living in a particular country? Once you have made your selection, can you get there by an indirect means, i.e. travel first to another country and then onwards to your final destination? How would you do this?

Do you need to change your name before you go? I would personally only change my first name and I would do all within my power not to let anyone know about it other than those who are absolutely necessary. Do you need to obtain a new passport and driver's license? Again, I would apply for a new passport in my new name while retaining one in my old name; I would also ask for a new driver's license.

Make a deception plan that really sounds true, and one that your family, friends, and neighbors will swallow hook, line, and sinker. For me, this is easy as I travel the planet on a regular basis anyway and no one, not even my wife, questions my movements. I could call her from Malaysia and say I was in New York, and nothing would be out of place—other than the time of course.

Timeline for Your Disappearance Exit

Unless your desire to disappear is immediate, then you should make a disappearance plan. Like everything in life, planning makes things happen in an organized way, and you will have more control over any situation. If you simply run out the door without thinking about where you will go or how you will live, your chances of success are limited. It is possible to leave almost immediately with very little money, but even this requires a rational and decisive transfer. Even walking out into the night and becoming another displaced person requires that you are warm, have somewhere to sleep, and food to eat. Additionally, simply walking out does not guarantee you will disappear for good.

In order to completely disappear and never be found, you need to plan in great detail. In some aspects, this might also give you time to reflect on the matter of your disappearance, as things can always change. If after all of your considerations you are still determined to start a new life, here is a basic disappearance plan.

Twelve Months Prior to Departure

- Decide you must and can really leave without having any regrets.
- Decide when and where you will go.
- Start a disappearance fund; harvest the money into a hidden stash.
- Start building your disappearance file; keep this in a secret place where no one will ever see it. Research as much as possible on all aspects of disappearing.
- Satisfy yourself that disappearing is possible, have no doubts.
- Determine how you will disappear, i.e. as a hobo, laborer, businessman, wealthy playboy, etc. This will be determined by your individual circumstances and personal wealth.

Nine Months Prior to Departure

- Decide where you will go: investigate all the possibilities and keep the information in your file.
- Decide what you will do when you get there. Will you join the Foreign Legion, work for an Nongovernmental Organization (NGO), or take a job overseas?
- Do you need to learn a language in order to fit in or work in your destination country?
- Do you need to learn new skills to live and work in your destination country?

- Start your deception plan, e.g., show interest in traveling, joining a NGO, and tell others of your hopes and ambitions.
- Check and increase your disappearance fund. Find extra temporary cash work or sell something.

Six Months Prior to Departure

- Intensify your deception plan: let family, friends, and work colleagues know that you really want to go help famine relief in Africa, have a desire to walk across Australia, etc.
- Visit the first country of your choice. If finances permit, take a short holiday to your country of destination without telling anyone. Once there, test out the local banking and work visa rules, living costs, etc.
- Obtain information on changing your name, applying for a passport, and getting a new driver's license. Keep the paperwork and forms hidden in your planning file.
- Plan how you intend to change your identity, if at all.
- Check and increase your disappearance fund.

Three Months Prior to Departure

- Adjust and refine your disappearance plan.
- Learn another language if required.
- Intensify your deception plan.
- Apply for an overseas work visa if required.
- Check everything is in place; apply for any forms you might need to change your name.
- Obtain a passport and driver's license in your new name.
- Check your financial position, convert some of your cash into prepaid credit cards, or buy gold coins. Secure your cash.
- Book and pay for your flight (if required); the earlier you do it, the cheaper it will be. Do it all online.

Now it's time to go.

Priorities in Your Disappearance Plan

Change your name. Keep this as secretive as possible and, although illegal to do so, continue with your old name until you disappear. Change only your given (first) name, say, from John to David, and add or remove a second given name, while retaining your family name (surname). Change your name on essential travel documents only, i.e. your passport and driver's license. If you live in America, move to a different state for a while in order to do this, obtain an international driver's license, and then move back to your original state. The idea is to finish up with a set of papers showing your new given name that no one knows about.

Travel on vacation to your preferred country of disappearance, but do not let anyone know you have gone. If you have the resources, travel there by a diversionary route. For example, you could say that you are going on business to Oman in the Middle East and book a flight accordingly. Once you have arrived in Muscat, buy a second ticket onwards to any country you like, such as Thailand, Indonesia, or Malaysia (I do this all the time—it's really easy). Use cash to buy the ticket and use the passport in your new name. Immigration in some of the smaller airports is merely a question of a quick look at your face before stamping your documents and letting you through.

> **Warning:** Don't forget, you will need to reverse the process in order to avoid any overstay flags coming up on the Omani immigration system. That means returning to Oman on your new name passport and then using your old name passport to travel back to the United States.

Visiting a Foreign Country

Don't waste your time sightseeing; get to work the moment you arrive. You will need to organize a bank account and set up a residence address. The latter is fairly easy in many countries, and a post office box address will suffice as a temporary measure. Be aware that

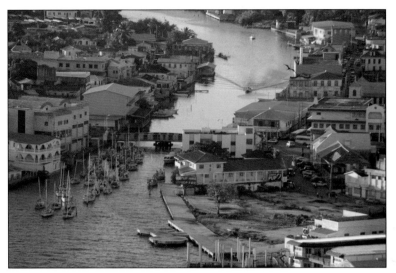

Once you have decided which country you plan to disappear to, you should first plan a vacation there. Belize City is a perfect place to disappear.

PO boxes will need emptying on a regular basis, so see if you can get one that does mail forwarding.

Depending on your country of choice, you will need to investigate what processes and documentation you require to set up a local bank account. This generally requires a residential address and a copy of your passport, but in my experience, banks are only too happy to take your money. If you can, set up an Internet banking system, whereby you can transfer money in and out from any location. Do not apply for credit cards until you have had the bank account for several months and have made at least two or three visits to the bank—always try and see the same individual at the bank, as this will ease the process. You can always make up some story that you intend to start a business at some stage and are just looking at various options. Just prior to your disappearance, you should visit your country of destination again and visit the bank; this time make inquiries about having a credit card. Once you have a valid credit card, you can safely transfer more money into this account. Personally, I would have a trail of at least three accounts in neighboring countries where I could get access to my funds, so that, if one were discovered, I could make use of the

It is always handy to be able to jump from one country to another. In Asia, the airline Air Asia flies to just about every country in the region for just a few dollars. For example, a flight from Singapore to Kuala Lumpur in Malaysia takes less than an hour and costs about $60.

others, transferring from one to the other.

Closing your credit card accounts at home before you disappear will only serve to highlight your intentions. If possible, you should contact your credit card companies and explain that you have a three-year contract overseas and will be away for some time—ask them to mark your account with the relevant information in order to avoid losing your credit rating. This will not be as obvious to anyone trying to locate you other than the government. Or you can simply say nothing and keep your credit card valid and normal until the moment you disappear.

It is possible to legally obtain two passports, both in the United States and Great Britain. The additional passport is generally issued to businessmen or people who frequently travel and need to send one passport to an embassy for visa approval because this can sometimes take weeks, during which time you are required to travel. Having two passports in the same name is a great opportunity to change your name in just one of them. There is always the chance that someone

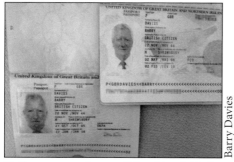

Both my passports are valid and legal. Notice one is issued overseas (FCO).

Barry Davies

at the passport office will notice you have been issued two passports, but this is unlikely.

If you have the means and the time, you can always obtain a second foreign passport from any number of countries simply because you have purchased a home or invested there. The trick here is to buy a home or invest and, once you have your passport, simply sell your investment.

Cover Story

When making a deception plan, you would be well advised to see how the military does it. The military takes actions to deliberately mislead enemy decision makers as to their capabilities, intentions, and operations, thereby causing the adversary to take specific actions (or inactions) that will contribute to the accomplishment of the friendly mission. This is called misinformation.

Misinformation methods designed to misguide and confuse the enemy are crucially important to the success of a government or military intelligence operation. The same process of misinformation can also be used when you plan your disappearance. This might involve deception at home, in your workplace, or within your social networks. But your deception plan must be consistent so everyone will get the same impression of whatever it is you're deceiving them about.

Disappearing overseas may depend on you taking work that is extremely hard. Becoming a lumberjack can help you disappear but it is one of the worst jobs in the world.

Telling everyone that you intend to go off and join the French Foreign Legion is not really a deception plan, although it could be part of one. The problem is, when you let people know that you intend to do something that is very different from your current lifestyle, you may also be giving an indication that you intend to actually disappear. Your plan should be subtle, secure, and only known to you.

So what constitutes a perfect disappearing plan? The first thing to do is keep everything looking and acting as normal; simply be yourself and tolerate whatever it is that is driving you to disappear and never be found again. In the first instance you must consider why you are doing this and who and what you will leave behind. You will need to make a list of all the points that will result in your successful disappearance.

Dropping off the Radar Slowly

Once you have finished your plan to disappear, start thinking about removing yourself from the radar; that is to say, slowly and carefully withdrawing any public information about yourself. For example, you might think it a good idea to start deleting any social network accounts such as Facebook or Twitter. However, when you do this, you leave a trail that shows some intent to disappear, i.e., people will wonder, "Why did they delete all their social network accounts?"

If you try to remove your account from Facebook, people will probably ask you why. Plus, it's not as straightforward as you think; all the comments you have left and photographs you have uploaded could still be attached to other people's pages. While it's not always easy to delete your profile from a social network, it is always possible to amend and edit your details. I would suggest you do this if you are having trouble removing it completely.

One of the best ways to drop below the radar is to let people simply forget you. In the case of family and friends, this is a bit difficult, but I know people who have done it. If your plan to disappear by serving in a foreign army, working for an NGO, or as a contract soldier, you might simply stay away from home long enough to be forgotten. If you are married, your wife will have found someone else, your kids will have grown up, and you will be nothing more than a memory. You can achieve the same result by going overseas to work in the construction industry, or as a mechanic in Libya; distance and time are the only two factors that matter.

Barry Davies

Working in the oil business generally means being in an isolated location.

Author's Note: I live in Spain and there is a young Iranian man who has lived for several years just down the road from me in a small, out of the way cottage. As far as I am aware he has no income, does not work, lives in the house free of charge, has no Spanish NIE, and for all intents and purposes does not exist on any radar. He is always clean and tidy, fairly polite, and keeps to himself—no one bothers him. He holds a British passport, so he is entitled to live in Spain under EU laws. He does not work so he's not required to pay tax; he is breaking no laws so the police are not chasing him, but he is living well below the radar.

Online Footprints

Most of the time, no one really cares about your online footprint; but when you go missing, they will start searching. The moment you log onto the Internet to either check your email or do an online search, you are leaving a lot of footprints. For example, while you are browsing web sites, "cookies" attach themselves to your computer. This allows the website to see how often you return. Search engines keep track of the items you have searched for, and they also keep track of the pages you have opened. In addition, there are lots of viruses, spyware, and adware lurking online, ready to attach themselves to your computer for a whole host of reasons. When you send or receive an email, you actually create a trail that connects you through various ISPs back to the sender or recipient, with your email's content being saved or even copied en route. Making online purchases requires you to upload your personal and financial details. The list is endless, and if you think your computer is secure, then think again. If you really want more control, beef up your browser security settings so your computer will warn you of uploads, downloads, and other attempted changes more frequently. You can also avoid some Internet footprints using an anonymous browser to hide your computer identity. While the websites you visit can still tell you are on the website, they can't tell where you're from.

Get some really good firewall software and browse the Internet in secrecy mode; use the best quality antivirus and antispyware software. If you want more privacy, consider a privacy software package, such as Privacy Guardian or Cyber Scrub, which can clean your memory and erase the footprints of online activity on your computer, such as email records.

Finally, you can mask your email identity by using an anonymous email account. I2P is an Invisible Internet Project software that allows applications to send secure messages and surf the Internet anonymously. Anonymous Internet tools like I2P are popular with those involved in nuisances and sending bulk spam, but these tools are also valuable to those who wish to protect their identity.

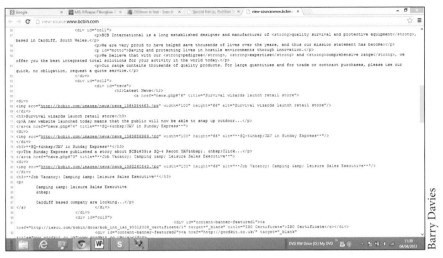

You would be surprised at what a little computer knowledge gets you when you want to see what's behind an email or website—so much more is revealed.

Email Accounts

Email is an essential part of life, and most people now use it instead of posting a letter. You will need to keep your email traffic to a minimum. If you are using email for leisure or business, then you will have a trail of whom you have communicated with and when you've done so. Do not stop using this form of communication, but be extremely careful of what you send out and what you attach. Never attach any recent images of yourself or indicate where you may have visited. Simply keep your emails direct and to the point, without any mention that could hint at your pending disappearance.

Web Searches

There will be times when you will need to access the Internet in order to research parts of your plan. You should do this in such a way as to not compromise your intentions or provide any indication that you are preparing to disappear. There are several different ways of doing this, the simplest being to use your planning dongle and an Internet café.

Barry Davies

Make sure your plan to disappear is kept secret—put it all on a removable drive.

Summary

Decide how you're going to disappear and what it will take to make it happen, i.e., do you require a new name and passport, should you tell anyone a deception or keep it secret? Clean up your past history or, at best, limit the amount of information available for others to find. Above all, plan out your disappearance step by step. The better the plan, the more you will see any faults and find a way around them. Finally and most importantly, keep your plan to disappear a secret.

The SAS have a saying, "Make sure your brain is engaged before opening your mouth." So before you even think about making a plan to disappear, let me say a word about talking to people. The only person you can trust not to say a word to anyone is YOU. You tell one family member or friend in confidence, and soon everyone will know—it's human nature. Never ever compromise your security. You must learn to be two people: the normal you who keeps talking about the same mundane things, and the other you with the secret plan to disappear. Keep the two completely separate, and never expose yourself to the risk of compromise by keeping your mouth shut.

5

COVER STORY AND SKILLS

When you are planning for your disappearance, you need to protect yourself against others observing you doing odd things. They might ask, "Why are you suddenly doing all these weekend markets?" or, "Why the sudden interest in walking every weekend and vacation?" Your answers should be as honest as possible. In the first case, you should reply something along the lines of, "I am saving for a round-the-world cruise and need all the money I can muster." In the second instance, you need only say that you want to get fit and walking in the great outdoors is your way of doing it. What you're actually doing is laying down a pattern of misinformation and ideas that will get planted in the minds of your family, friends, and coworkers. The list of disinformation you can use is never ending, and every false alley will help you achieve your final goal.

However, when it comes to your immediate family—especially your spouse—you are going to have to be very careful about how you substantiate your actions and any money you spend. Why would you want to go on a world cruise without taking your family with you? Why would you go camping most free weekends on your own? Plus, you would never mention to your wife that the girl of your dreams is living in Thailand. The solution is to tailor your cover story to suit your individual situation: If you're married with children, single, or in a long–term relationship, all will need a cover story that is 100 percent plausible. If you are single and living on your own, then you have few worries; if you are married or living with someone, anything you do out of the ordinary is going to cause suspicion.

Remember, you are going to do a few strange and impulsive things, so think about how you are going to answer your questions when asked why. Here are a few examples:

- I am learning Spanish!
  Your answer to why could be any number of legitimate reasons: You live in a part of America where Spanish is widely spoken, you are thinking of taking your family to Peru on holiday to see Machu Picchu, or maybe your company is looking to move into the South American market.

If you can ski, you can always find work as a skiing instructor.

- I am going on a world cruise!

  This is okay if you're single, but if you're married your family will be expecting to come along, too. Other excuses could be that you are going on a volunteering mission abroad for which you have to pay your own way. Your family is less likely to want to join you in some ghetto in South America.

Going on a survival course is a great idea as you will learn a lot of very useful skills, especially if you intend to disappear as a hobo.

- I am going on a survival course; I fear that doomsday could be upon us!

  If what we see on UK television is true, then lots of Americans believe and practice this—learning how to shoot, stockpiling food, etc. is the norm. It makes no matter if you're married or not, as you can always invite your family to join you. If you are considering this as part of your disappearance plan, most likely the situation at home will be so bad they will not want anything to do with you anyway.

- I go away each weekend to learn how to sail!

  I have always dreamed about one day sailing around the world. Again, you can invite your family to join you; chances are they will not be interested because you tell them the water is rough and they will get sea sick. If you learn the skill of sailing and continue to talk about your desire, when you disappear people will think you are simply accomplishing your dream.

- I have always wanted to marry a South American woman or man and have been looking on some dating websites!

  This one is mainly for the single people thinking of disappearing and not one you should share with your wife or husband. You might build the cover story by actually renting a South American escort and having your friends see you.

My point is that all the reasons above can be explained away as a legitimate cover for your actions. You will need to be realistic, as there is no point in a married man of, say, fifty telling his wife that he intends to run off and join the French Foreign Legion. You will need to keep your cover story as accurate and truthful as possible. Build on it, but keep it factual and practical.

The first major problem you have is deciding whether you need a cover story or not. You may well already possess the required skills: you may already speak Spanish or Hindi or hold a sailing certificate that enables you to sail single–handedly around the world; you might be thirty-five years old and recently retired from the military. As part of your plan to disappear, you may need to make up a story that justifies your eventual disappearance. Nonetheless, in my humble opinion, if you can disappear without having to make up misinformation, you are better off avoiding it. The problem is that the more you say, the more you raise suspicion. In the end it will reveal that you planned to disappear in the first place. You may well choose just to carry on as if nothing out of the ordinary is happening until the day you go, and it will be only then that people will suddenly realize you have disappeared for no apparent reason.

On the other hand, disinformation does offer you a means of covering your actions should they be discovered, and it provides a legitimate reason for you learning new skills.

Sample Cover Story

You are forty-eight years old and feel your age. You are married to a woman who drinks far too much and cheats on you whenever she meets someone new at a bar. Luckily, your only daughter has grown up, recently gotten married, and set up home with a great guy. You work all week for a large building company—a profession you have done since leaving high school, but your weekends are mainly spent alone as your wife is off visiting her sister (or so she says), and when she is home she is verbally abusive, with alcohol playing a major part.

The construction work in Abu Dhabi is never ending and employment opportunities are always extremely good.

You wake up one Saturday morning and see a program on the television about construction work in Abu Dhabi, during which there is mention of a shortage of skilled building managers. You realize that you could do that job standing on your head and wish you were there. Back at work on Monday morning, you discuss the television program with one of your colleagues and he tells you of a friend of his that went out to the Middle East as a building contractor and made a lot of money. The problem was he was gone so long his wife divorced him and took up with a local football player. For the rest of the day you dwell on the story—that could have been you. The seed is sown, and you develop a plan to disappear. For the next few weeks, you work on how you're going to start a new life.

On one of the occasions when your wife is home and relatively sober, you breach the subject of you working overseas. You say that later in the year your company will possibly lay you off due to the decline in the local economy. What you are doing is instilling the fear of financial loss within her. Now your wife's main focus will be on her home and your income. Almost immediately, you provide a solution and save the family home: you're planning to work overseas in Abu Dhabi. You have around six months to prepare, in which time you must seek out a good Middle East employer and learn some Arabic. You will also need to seek a work permit, visa, and so on. In essence, you have to explain to your

For those who have never traveled and worry about going overseas, let me tell you Abu Dhabi City is one of richest in the world—it has great hotels, beaches, and bars.

wife that you will be away for months at a time, and you hope she can cope with this. The fact that she will remain in her home and you will be earning good money, which she assumes you will be putting directly into your shared bank account, makes her extremely happy. The fact that you're away in the Middle East working means she does not have to hide her infidelity—she is extremely supportive of your plan and even offers to visit you in Abu Dhabi from time to time.

Now you are free to plan your disappearance. You might as well learn Arabic (it's easier than most languages) and do some research on the Internet to see what actual employment is available in Abu Dhabi. You talk to your colleagues about the idea and manage to get a couple contact names of guys already working out in the Middle East.

Now all this will work, and you may well get employment and earn a good salary. Your work contract may be for three years and extendable; you work hard, do an excellent job, and people like you. Now let's stop and think: How much of your salary do you send home to America each month? All of it, or just enough to cover the mortgage, pay the utility bills, and keep your wife in alcohol? Remember, you do need to live in Abu Dhabi (the most vibrant city in the Middle East), and it's not cheap. However, you have discovered that you can save around $1,500 per month and then there's the terminal bonus of $15,000 over a period of three years. You could save a nest egg of close to $70,000.

Your infrequent visits home are not happy. Your wife, by this time, has almost had her latest boyfriend living in your home and sleeping in your bed. The first thing she asks is how long you're staying, quickly followed by what happened to all the extra money you were supposed to earn. You explain that 30 percent of your salary is held in reserve and will be paid in a lump sum when the three year contract is up. It's a lie, but she is not going to check.

Slowly but surely, each time you are home, you spend most of the time selling off old items you no longer need on eBay or at a local market. Items that are personally important to you are slowly put into storage, ready for shipment later. There are lots of storage companies that don't ask questions, and any shipping agent will ship your goods to just about any place on earth; it will be just one more consignment and virtually untraceable.

When the time comes, just don't bother coming home any more. Stop all the money transfers, and move to work in another country without telling anyone.

Author's Note: While the sample cover story is made up, I would like to relate to you a real one that is similar. A close friend of mine, Alan, left the British military (SAS) and got a great job as bodyguard for the Saudi Arabian Oil Minister, Ahmed Zaki Yamani. The job was extremely well paid and, like a good husband, he sent all his money back home. One day, he arrived unexpectedly and found that his wife had moved in her lover, a member of the Hereford Football Club. Finding this guy in his home, he confronted his wife who basically said take it or leave it. That night Alan had to sleep in a hotel. He also had a young fourteen-year-old son whom he loved very much, and it pained him to see his son unhappy.

The next day, he made arrangements to quit his highly paid bodyguard job, as he wanted to salvage his marriage for his son's sake. The reply he got from Yamani was, "No one leaves my employment, but I understand your position and have made arrangements

Sheik Yamani

for you to look after my property in Switzerland." Alan went to Switzerland and found the property, a beautiful place in Chermignon. The property was spacious, had several top-of-the-range sports cars, and a beautiful young Swiss girl as housekeeper. Alan took up Yamani's offer and moved to Switzerland. Several weeks later, he turned up at his house in Hereford driving a Ferrari with the glamorous housekeeper sitting next to him. He collected his son and drove off into the sunset to start a new life.

Skills

Disclaimer: Unless carried out as part of your professional trade, to go about in public carrying lock-picking tools, breaking into houses, or hacking someone's mobile phone or computer is illegal. Mention of these skills here is solely in the context of techniques that have been used many times by others and that are freely available in the public domain. Also, some readers might construe the information written below as irresponsible. I would like to assure the reader that it took me two years to learn the very basics, and that was with constant practice and expert instruction. Moreover, if a person wishes to learn such skills, there are plenty of websites to show them how. Therefore, I take absolutely no responsibility for anyone learning or using the skills mentioned. Neither do I encourage you to break the laws in your country.

Most of these skills are handy, but they should be used within the context of purpose. Imagine you have become a drifter and one bitterly cold winter's evening you stumble across a small summer home in the forest. Who could chastise you for breaking in and seeking warm, dry shelter? If you know how to pick a lock, it will save you from having to smash a window. Should you find yourself in such a situation, always leave the place in the same condition as you found it, and all will be well.

It makes no difference if you're going to disappear to a foreign country and start a new life or simply become a drifter or hobo in your own. There are a number of skills that will help you a lot. You might believe that many of these skills are not necessary and you could be right, but believe me, they are all worth learning. The skills are all different and they are in no logical order.

You never know when you will need a few breaking and entering skills.

Breaking and Entering

During my lifetime, I have broken into many houses, offices, and factories, all in the name of military and counter–terrorist operations.

Many were in built up areas, but my advice to you would be to ignore homes in cities, towns, or villages. You may need to break into a remote home if you are disappearing as a hobo in order to get shelter, or steal food or clothing to keep you warm.

There were a few basic rules that I needed to apply, such as doing some reconnaissance before approaching the target home. For example, is there any washing on the line? If the answer is yes, then the home is occupied. Likewise, if you spot a home that has a wall or fence around it, simply put a small stone on the gateway in such a manner that if the gate is opened the stone will fall off. This way, over the next few days, you can detect if the property is in use or not.

If you want to make sure a home is unoccupied, ring the front door bell or knock aggressively (don't rely on ringing the doorbell because it might not work). Make it a long and loud knock, just

Barry Davies

If you must smash a window to gain entry, always choose a small pane close to the window catch. Tape the glass beforehand to prevent injury and reduce noise.

in case someone is sleeping in a rear bedroom. If someone should answer the door, you need to have a ready response. You might ask for a fictitious person or directions to a nearby town. If no one comes to the door after repeated knocks, you should proceed to the rear of the house and attempt the break-in.

Once you have entered, the first thing to do is make sure you have several quick escape routes. To do this, go to the front, rear, and any side door to make sure they are unlocked. If the doors are locked, open several windows wide enough to escape; be careful on a windy day, as doors can slam and cause unnecessary noise. Take what you need—food, clothing, cooking utensils, etc.—and leave. Do not trash the home or steal items to sell. Remember, you are not a thief; you are just looking to try to live (find food) or improve your situation (find shelter).

Learn a Language

I have mentioned several times in this book the significance of learning a language, and I cannot stress its importance enough if you plan to disappear overseas. While English is the most widely spoken language in world, it ranks third in popularity to Mandarin Chinese and Spanish. The language you learn should be tailored to your country of destination once you have disappeared. For example, if you're going down to South America, then Spanish would help you most, or if you were going to Asia you might decide to learn Malaysian (Bahasa Malaysia), as this is fairly compatible with Indonesian (both countries speak a lot of English).

By far, the best way to learn a foreign language is to visit the country and plunge yourself into its culture and language, but this is rarely possible. Today most people use computer software, such as Rosetta Stone (rosettastone.com), which is one of the better language learning systems. Additionally, there are many simple apps for your smart phone that will provide a great vocabulary.

Author's Note: During my life, I have learned German, Norwegian, Spanish, Arabic, and some Malay. At the moment I am trying my lick with Mandarin Chinese. My German is still quite good, but you need to speak a language you learn constantly in order to be proficient. I have found

Barry Davies

Banding tape is easily slipped inside most car windows or door frames.

that vocabulary is the key to quickly learning a language. Learn as many words as you can—forget about the grammar. If you know the words, the rest will fall into place—maybe not the right place, but enough for people to understand you.

Breaking into a Car

Breaking into a car does not necessarily mean you're going to steal it; you may want to break into your own car having lost or locked the keys inside. If you have lost your keys and need to drive to town to get a new set, then most vehicles can be started without an ignition key (apart from the very newest models that no longer require a key to start the engine). So if you are thinking of opening and hot-wiring your car, then I suggest you try the following if it is at least five years old.

Gaining access to your vehicle is fairly simple, and there are many ways of doing this. My favorite method that will work on most older models is to use a piece of banding tape, the kind you find holding large cartons together. You can find a one meter (three foot) length in most trash cans on industrial sites or commercial premises. You simply fold the tape in half and slide it into the top outer corner of the window (the weakest point). Next, use a sawing motion to pull it down close to the door lock and push one end of the tape forward so that it forms a bow. Hoop the bow over the door lock and pull using a backward and upwards force. Your car door should be open without having to smash a window.

Barry Davies

Pushing one side of the banding tape makes a hoop that fits over the door lock.

Some people will tell you that it is possible to open a car door by using a tennis ball. The idea is that you make a small hole in the tennis ball and set this close to the keyhole on the door lock. You then hit the tennis ball with force and the air will jump the lock open. I have tried this method about 50 times on a variety of cars and it has not worked for me.

Once inside, you will need to start your car. Look under the dash for two red wires (in older cars red was the standard color). If they are not present, look for two that make a matching pair. When you find them, cross them together and take off! A word of warning, some cars have a well-protected casing around the wheel column that you might have to remove first. The thing to remember is that there are lots of wires running down the steering column. The ones you need to cross are the two that are the same color. There are also other ways to start a car, such as using a wire from the battery's positive terminal to the coil and jump-starting the engine that way. Alternatively, in an emergency where you need to start your car in a hurry, you could try putting a flat–head screwdriver into the key slot, twist to break the lock and start the car (works most times but not always), you will have to buy a new lock once you have it started.

The only thing you need to be aware of is the vehicle alarm, as most have one fitted. This will activate after a few seconds and will continue to run until the engine is started, so you might want to warn the neighbors. You will need to do some research—there are lots of Internet sites that show you how to get your car running if you have lost your keys.

Garbology

Whether you are looking for food or planning on stealing someone's identity in order to disappear under a different name—bins and dumpsters are a haven. If you're looking for food, restaurants and larger homes should be at the top of your list. Check if it's good enough to eat, and be on your way.

However, the main reason for learning garbology skills is to steal information on someone. To do this, you need to build up a file on the target person and gain as much information as possible about them. On the other hand, you might use your skills in garbology to protect yourself once you have left home for good.

Garbology simply means taking the trash or garbage from a household or business and examining it. It can prove to be a very good source of information, providing confirmation and relevant details about a target. The idea is to collect a target's garbage discreetly and

Barry Davies

Garbology shows that this bin has both receipts and medication used by someone in the home.

examine it at your leisure, making a detailed record of all items. The main advantage of garbology is that it is unobtrusive and almost always goes unnoticed. You will be surprised at the amount of information you can find.

Never throw anything into the garbage that can prove your identity or provide important information to others. Buy a steel bucket and make a burn bin. Simply burn all your unsolicited mail, unwanted bank statements, etc. before disposal. Only put discarded food in the garbage bin, and take bottles to the bottle bank. If you think you are being watched, discreetly place your garbage in someone else's bin. Never throw away any statements that have address correction slips attached, as a person can request a new card be sent to a different address, and they may already have a signature from discarded credit card receipts.

If you want to search someone's trash, watch and observe the best time to collect the target's garbage. Garbage disposal systems vary from town to town and country to country, but the final pick up by the garbage collector is normally inevitable. Watch and make a note of the garbage truck's date and time of arrival at the target's premises. Check to see if the garbage container is for individual use or multiple households. In many countries, the individual must take their garbage to a shared container; in this case you must establish the target's habit for taking out their garbage.

The hours of darkness are best to collect the garbage for searching. However, if a daylight pick up is required, dress accordingly—such a hobo—and carry a plastic bag. Remember to wear a pair of rubber gloves, as some bins can be unhygienic.

Lay out the garbage contents on a large plastic sheet and discard all useless items, such as food waste. However, always keep notes as to the type of food being consumed, i.e. fast food or expensive food. Next, check each individual item and make notes on each. For example:

- Count the number of cigarette butts and identify the brand.
- Count the number of alcohol bottles or cans. Identify the type and brand.
- Set aside all correspondence, including papers such as telephone bills and bank statements, for detailed examination later.
- Photograph any items that might be of interest, such as discarded clothing, magazines, computer disks, and empty nonfood packaging.
- Carry out an in-depth examination of all correspondence and paper products.
- Write down your observations and conclusions for each separate garbage pick up. You should carry out at least four separate pick ups within the period of a month in order to make a minimum assessment.
- Record any important discoveries you think may be of immediate interest.

The outcome from a good garbology probe over several weeks can be very revealing. The information gathered should be properly documented with a list of attributes added to the target's file. Some examples:

- Correct name and address (all correspondence).
- Personal finances, including the name of the bank and account number (statements).
- Credit card usage (statements).
- Target's signature (discarded credit card receipts).
- Telephone numbers, especially repetitive numbers (itemized telephone bills).
- Email addresses (letter or discarded printed email).
- Work or employment address (pay slips).
- The amount of cigarettes smoked (cigarette butts).
- The amount of alcohol consumed (amount of bottles and cans).

- Known toiletries (discarded bathroom waste).
- Rough weekly expenditure (total of consumed items' original costs).

The list of clear and precise information that can be obtained through garbology is endless, but it needs to be done in a methodical way and there are some things that will need to be taken into account. One factor to consider is how many people are being catered for at the target's premises. If the target lives alone, this is not a problem. However, one quick way of determining household numbers is by carrying out a clothesline assessment.

Avoiding Dogs

Dogs are a real pain in the butt when it comes to breaking into other people's houses or premises. Unlike an alarm system, they have teeth and speed. Should you find yourself being threatened by a dog, your smartest move is to back off. First off, its bark will alert any neighbors, and if you do gain access to the home, the dog will most probably attack you.

Should you ever find yourself being chased by an angry dog, see if it's possible to find a corner of a building or a tree to hide behind. The dog will have to slow down or stop in order to confront you. This is your moment. Kick or hit the dog once it is stationary; no matter how large it is, any normal human can defeat a dog. It's only fear that stops you. If it has a collar, grab and twist to

Barry Davies

The clothesline is a wealth of information when it comes to identifying how many people live in a house and their gender.

cut off the dog's air supply, and lift it into the air at the same time. Hitting a dog hard on the back of their spine will cause a lot of damage. I have also found that a charging dog can be distracted by simply charging directly towards it; put your arms out wide and scream at the top of your voice as if you were the attacker. Dogs are not stupid and will shy from a larger animal if they think it will harm them.

Hack a Computer Password

Access to someone's computer can reveal a lot about a person and, with luck, can supply you with a list of passwords (most people keep one on their computer). There are so many Internet sites explaining how to do this, and one good starting place is http://www.computerhope.com/issues/ch000806.htm.

Hack a Mobile Phone

How often have you found one of your old mobile phones but have forgotten the password? Not to worry, once again there are lots of Internet sites that will show you how to hack a mobile phone password. As there are so many different makes and models, I have simply chosen a common one: the iPhone 4. The latest software update (from when this work was written), iOS 6.1, appears to weak, as it lets you carry out the following: youtube.com/watch?v=-uCZWiUdfv0.

How to Pick a Lock

Most locks that have been manufactured over the past twenty years are of the pin tumbler type. In its basic form, it is a simple locking device. A series of small pins fits into the inner barrel of a cylinder. The pins are split in the middle, normally at different lengths, and are forced into recesses within the inner barrel by a small spring. If the correct key is inserted, the different sized pins are brought into line where their split meets the outer casing of the inner barrel. This

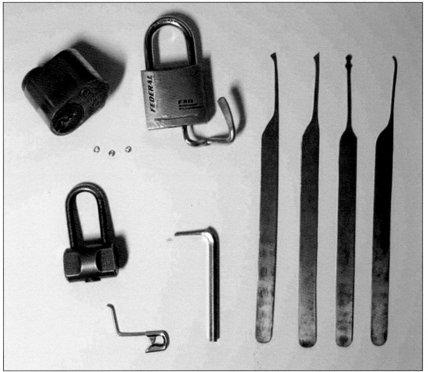

Barry Davies

Lock picking is a great skill to learn.

allows the inner barrel to turn freely within the casing, thus releasing the lock.

Any method of aligning the pins in this manner and turning the inner barrel will open the lock. This can be achieved by racking and picking the pins.

The tools for such work are not normally available in most countries, but it is fairly easy to manufacture homemade ones from a set of mechanic's feeler gauges. Two basic bits are required: a lock pick, or rake, and a tension bar. The pick is a flat strip of hardened metal that has its end shaped to fit into the lock and advance the pins on their small springs to the required depth. The tension bar is a simple flat strip of metal, inserted into the mouth of the barrel to employ a minute amount of tension. This process helps to seat the pins and turn the barrel.

Author's Note: There are many different types and designs of lock picking tools, and they all have different functions. I would suggest that the two mentioned are sufficient.

Raking is the quickest method of opening a lock as it is fast and easily done, providing the pins are not mushroom shaped, which are top heavy and can bind the lock. The lock should be clean and free from any grit or dirt; blowing hard into the lock before attempting to open it is a good idea. Raking is simply a matter of inserting the pick to the rear of the pins and swiftly snapping the pick outward, running the tip over the pins in the process.

Prior to doing this, the tension bar is inserted into the bottom of the keyway and a slight pressure is applied on the lock's inner barrel. The tension should be applied in the unlock direction and should be just enough to turn the barrel once the pins are seated, but not so strong as to bind the pins against the barrel. It is this single "feel" that is the basis of all good lock picking. If the tension is too heavy, the top pins will bind and the sear line will not allow the breaking point to meet. If tension is too weak, the pins will simply fall back into the locked position.

When raking a lock, it will be necessary to repeat the operation several times. If the barrel does not turn by the fourth time, hold the tension in place with the tool. Place your ear to the lock and slowly release the tension; you will hear the pitting sound as the pins fall back into place. Try to count the pits, as this will give you an indication of how close you are to opening the lock and whether you need to individually seat some of the pins.

Seating individual pins is similar to raking, but requires a lot more skill. Staring at the back of the lock, feel for the rearmost pin and gently push it up. The barrel should move a minute fraction. Working towards the end of the lock seat, do the same for each pin in turn until the barrel is released. Experience has taught me that a combination of one swift rake followed by individual pin picking is sometimes the best answer.

Rapid Change of Identity

Once you have disappeared, if you have valid reasons to believe a skip tracer could track you down, you will need to take precautions. If you think you are being followed, one way to throw any tail off your back is to have a rapid change of identity. This is also a good ploy if you have gotten into trouble and want to avoid recognition. People always describe troublemakers by the color of their skin and what they are wearing. There is very little you can do about changing your racial origin, but there is a lot you can do about your appearance.

Buy outer clothes that are reversible, providing a distinct color change. Carry a baseball cap, tie, or plastic raincoat in your pocket. Walk into a pub or any place that is heavily populated, change as you move through the crowd, and leave at another exit. If you were smoking a cigarette when you entered, discard it before you leave.

Barry Davies

A baseball cap and a set of sunglasses make for a quick face change.

Keep a folded shopping bag in your pocket, and carry this into the pub, but leave without it.

One way to lose a foot tail is during the rush hour on public transport. Catch a bus or train and always sit (or stand) close to the door. When you get on, see who follows. Stay on for a long time and see if any of them get off, or watch the timing of the doors. You can usually predict the exact moment the doors will close and then you can make a dash for it.

Summary

There are many different ways to openly preplan your disappearance and have a good cover story to justify it. For example, you might say you are traveling around the world; after six months only your close family and friends will worry why no one has heard from you. If your last port of call was in South America, you could have been kidnapped, or even killed, and your body never found. If you traveled to Australia, you might have become lost out in the bush. Chances are your body will never be found.

You may have told everyone that you plan to join the French Foreign Legion and might actually do it for a few years, or you have volunteered to join a NGO and travel to Africa to help underprivileged refugees. Any one of these excuses is viable, and you can disappear when you are ready.

Building your cover story requires serious attention to detail, as you will provide family, friends, and work colleagues a fabrication of lies. Make sure that the cover story that aids your disappearance is one that raises no suspicion. Hone and develop the skills you think are appropriate for your disappearance. If you intend to go overseas, learn the language. If your new life requires you to be hobo, make sure you know how to keep warm in the winter.

6

PLACES TO DISAPPEAR

There are many ways to disappear and many places to go, but as I have said previously, most of them require a lot of planning. In order to survive in your new life, you will need to decide what you will do when you disappear. You may simply decide to walk out and keep going, to become a tramp or as the Americans say, a hobo or drifter. On the other hand, you may decide to join a foreign army or become a monk in some faraway monastery. No matter what you decide, you must always think it through meticulously.

Disappearing is not just leaving home and never being seen again; it's leaving one life behind and starting a new one. No one simply walks into a new life; there has to be some planning and preparation beforehand. Additionally, you will need a set of skills to start your new life. For example, if you are to be a hobo or tramp, then you will need to know how to live rough; if you intend to join foreign army, you would be well advised to have some military skills. One of the basic skills you will need is to speak the language of the country you intend to live in, especially if that language is different from what you speak now.

When we were young, the older generation would ask us what we wanted to be when we grew up. The joke was always to say you were going to run away and join the circus. As silly as it sounds, it is possible to do just that, but in doing so you have to make a plan—a plan that involves having some talent that is applicable to working in a circus. You won't be accepted without one.

Now there is a whole range of possibilities when it comes to disappearing, and some are more pleasant than others. The trick is to match your method of disappearance to your skills. It's com-

Hobo life in the city is not good. Even if you manage to find an organized place to rest your head, you are still not safe.

mon sense, just like being in the military: if you're a very good shot then you go to Sniper Division. The same rule applies—do what you're best at—so if you're good with electronics then become an IT employee. To help get your thoughts in order here is a simple list of just some of the possibilities.

Hobo

This is the simplest and most affordable form of disappearing; all you need to do is walk out the door and never come back. The drawback to being a hobo is that it's a really hard life (see Chapters Eight and Nine). You need no skills and very little money, but you will have to live on your wits and know how to survive.

Why would anyone want to live rough? Living rough generally means you are a nonentity as far as others are concerned. Most people will try to avoid you, and if you get any problems at all it will normally be from the police moving you on. Dropping down into the lower end of society has many advantages. It means you can avoid much of society's red tape; you can beg, you can borrow stuff people leave lying around, and in most cases, you are free to travel within the confines of a country.

There are various stages of being a tramp or hobo. For example, you can be a real tramp and dress in stinky old clothes, never shave, never wash, and simply go around living rough. On the other hand, you could keep yourself fairly tidy, clean, and well-mannered, so that while you may look a little jaded, you will still be acceptable for work or odd jobs. One way of looking at being a hobo is to consider yourself a long-term camper carrying your tent and sleeping bag, spare clothes, utensils, and everything you need in your rucksack. When you think about it, many of the American forefathers were trappers and woodsmen who lived in the same way as a hobo does today. They carried most of their possessions with them, had a regular set of shelters, and simply lived off the land.

When you are out on the road driving an RV, you can stop where you like, you have a permanent bed, place to cook, shower, and most of all you can keep to yourself.

Overland Drive

You could purchase or rent an RV and tour across America and beyond if you have the right paperwork and your rental allows it. An RV will provide you with a bed, cooking facilities, and the means of getting wherever you want (on land). The drawback is that it's very expensive, as the average rental cost is between $100 to $600 per day depending on the size of RV.

The great thing about RV travel is that you are constantly on the move—or not, depending upon what you decide. You might find a great location by a lake and stay there for a few weeks. You can hide your RV away in the wilderness or you can drive into the city, as the RV offers a lot of freedom. It is certainly a good idea for those people who want to disappear but are still not convinced it's the right thing to do.

In the UK, we have the Caravan Club, membership of which provides benefits such as special campsites; these are very much the same as RV Clubs in America. However, there is no law that states you have to join a club, and renting an RV is really simple.

To prove the point, I went online from here in Spain and inquired about booking an RV vacation in America. All they required was a full British driver's license and a credit card (Visa or MasterCard). They also wanted to know my age. I found a number of well-established companies offering booking via the Internet, with elmonterv.com and roadbearrv.com being just two of the possibilities.

As the RV was just for one person (I said it was for my wife and I to allay any suspicion), I decided on a modest 25 footer as opposed to the lager 40-foot model, because I thought his would be cheaper and easier to drive around and park. Next I investigated places to stay, and I must say that America has an incredible network of beautiful places in both State and National Parks. Most provided basic power hookups shower facilities, laundry, and even Internet access. All of this makes disappearing sound very attractive. The great thing is no one really knows who you are—you are who you want to be!

Providing you do not get stopped for speeding or any other misdemeanor, there is no reason why the police should pull you over. Even if you do get stopped, you will have your driver's license and the vehicle's registration papers: who's to know that you left home in Spain simply to disappear? There will be

Working for an NGO in Syria is possibly not the best place to disappear, but it is a great way to hide.

limits due to financial constraints on how long you can rent an RV, but if you have the money you could always buy your own.

With your own RV, you could travel the length and breadth of the North and South American continent. Likewise, in Europe you could drive from the UK as far afield as Portugal or Greece without being stopped (there are occasional checks at some border crossings). To supplement your travel costs, you could always look for work or do whatever you do to earn money. The only downside is that RV or Caravan travel costs money, not just the fuel but campsites, vehicle maintenance, and so on.

Join an NGO

Joining an NGO will take a little time, and if this is the route you have chosen to disappear, then I suggest you start your planning at least eighteen months in advance. There is no real downside to joining an NGO, other than that they usually operate in the most awful places. If you choose to go down this path, you really need to do some research into what the NGO does and where. If you go to ngo.org/links/list.htm, you will get a list of NGOs that are affiliated with the United Nations. You can check out those that might be suitable for you to join.

Now it is only fair to warn you that joining an NGO to work overseas is not as simple as it sounds. In the first instance, most of your initial inquiries will involve a discussion on how much money you would like to donate or what you can do for the organization, which is more about raising money or organization awareness. Your best bet is to go for an NGO that supplies food to areas where there is famine.

Despite politics, NGOs often push for food aid when and where it is most needed. Because NGOs operate at the ground level during emergencies, they are first-hand witnesses of how food aid is used and when it is and is not needed. This means they need teams of dedicated people who will travel with the food supplies to make sure they get to the right people. So my advice would be to investigate how you get to be a member of such a team.

As a place to start, and to give you some ideas you can go to USAID, which will then transfer you to devis.com. There, you can look up careers. As I said, while working for an NGO is a great idea, it's not easy. Remember that your end game is to disappear, and joining an NGO is only the first step in getting you overseas with a justifiable reason.

Run Away to Sea

Sailing or getting on a ship and running away to sea are also great ways of disappearing. You have lots of choices and very few restrictions on where you can go. There is really no downside to this form of disappearance unless you plan to sail away alone.

The great thing about running away to sea is that you can go just about anywhere. However, bear in mind that water covers 71 percent of the earth's surface, and that's a lot of water! I have a problem with disappearing at sea, as man is not designed to live in water—we are a land–based mammal. Even so, it is an overriding fact that the sea is the perfect place to disappear and never be found as, if you fall overboard, that's precisely what will happen. Your choices when considering going to sea are basically to go on a cruise and disappear, join a ship and work your passage, or get your own boat and just take off.

If you plan it right, you could always book a world cruise and disappear for some legitimate reason at a port of your choice. Let's say that after two months, your world cruise ship pulled in at the Bay of Islands in New Zealand. This beautiful place is where the Treaty of Waitangi was signed, establishing British rule and granting the native inhabitants equal status—a great place to live out your days. You simply go ashore and walk off into the sunset. You will need to hide yourself away for a period of time, but camping out there is easy. True, you would have to plan it in more detail, but at least you will have started your disappearance by being at the other end of the world.

Because world cruises are not cheap, you might also consider working on a cruise ship. There are almost as many staff members on a good cruise ship as there are passengers. While the majority of work is ship related (engineers, etc.), there is also a great demand for catering staff, chefs, and the like. Then there is the entertainment and bar staff. Most major cruise ships have a very high standard, so you will need some qualifications.

Finally, you could always buy or lease your own boat and sail around the world at your leisure. While this might seem to be a very easy and logical approach to disappearing, you will need to have some sea-worthy training. You will need to be good at navigation and have a good craft; one that is easy for a single person to sail and one that will actually get you around the world. You're going to need complete proficiency in sailing and seamanship together with a high degree of self-sufficiency. Handling a sailboat single-handedly will

Running away to sea will certainly convince people that you have disappeared, but you will have to touch landfall at some point, which can cause problems.

also require physical fitness in order to make sail adjustments and changes, such as wrestling the jib down and off the foredeck in a sudden storm. You will also need a boat that is equipped to sail the oceans of the world such as a recreational Westsail 32 round-the-world yacht.

So while simply disappearing out to sea looks appealing, unless you have the aptitude and sailing experience, I would look elsewhere. I personally don't think I could live with the thought of falling over-board in the middle of nowhere, with no one to even hear me, let alone rescue me.

Conversely, going to sea is not the only place to live on water. One of the most pleasurable ways to disappear—either in the short- or long-term—is to take to the inland water. The landmass of North America is dotted with huge lakes and wide rivers, and it is simple

The inland waterways where it is possible to live on a small canal boat exist around the world. These include America, Canada, the United Kingdom, France, Germany, Holland, and Russia. Just like an RV, you can live happily on a canal boat without anyone finding you. No highway patrols, no permanent residence, no one to bother you.

enough to buy or rent a houseboat and live in quiet, isolated tranquility. In the United Kingdom, there is a whole network of canals that were constructed during the Industrial Revolution. Today, these canals form part of a vast inland pleasure park where people take a holiday or live in residence on a canal barge. This offers those wishing to disappear into a more peaceful and tranquil life the ideal setting. Living on a narrow boat on the canal is perhaps one of the best ways a person could simply disappear. You are in the middle of civilization, but apart because you are on the canal. The chances of anyone spotting you are fairly remote, and even if they do, no one takes any notice of a barge on the canal.

Join a Foreign Army

Joining a foreign army is not so much about disappearing but more about moving away and being forgotten for a while. This has many advantages, as it will let your family and friends all believe you are alive and safe. The drawback here is the discipline and the fact that you might end up fighting in some remote corner of the world and get killed.

The traditional way for men to disappear was to join the French Foreign Legion. This was the last option of desperate men trying to escape any number of crimes, debts, or bad love affairs. Today, it is still possible to join the French Foreign Legion, however the rules have changed somewhat.

The French Foreign Legion is nothing like its portrayal on the cinema screen, where the men are defending some Beau Geste fort against overwhelming odds. The modern Legion is a professional fighting force with an excellent reputation around the world. This elite fighting force draws men—there are no women allowed—from all corners of the world.

Formed in 1831, the original Legion acquired a reputation for being a haven for cutthroats, crooks, and fugitives from justice. Few questions were asked of new recruits, making it an ideal repository for the scum of the earth, jilted romantics, and men searching to dull the pain of a lost love. Today, this Hollywood version of the French Foreign Legion could not be further from the truth.

The Legion today is 7,500 strong and still operates around the world. While most of its officers are French born, the majority of the other ranks are enlisted from outside France. So what does joining the Legion have to offer? The first thing is a lot of hard training, and if you've never been in the military before, you're in for a shock. The recruit training is severe, and you will have to endure a fair amount of punching and kicking, as well as learning basic military skills. All recruits have to speak French, and if you don't already, then you have to learn. This includes an awful lot of swear words, which seem to make up most of the vocabulary in the Legion. The Legion still drives its trainees to scrub floors manically, fold kit and uniforms with obsessive precision, and also do a lot of marching.

So if you plan to disappear into the French Foreign Legion, there are a few things you should remember. They do not take murderers anymore—you must provide your real name, and you must be willing to take the punishing training without complaint. On the upside, you will make friends in the Legion that will last you the rest of your life, find travel and excitement, and feel a part of something. You will also receive French citizenship should you wish it, and you will be well paid.

Author's Note: During my military service with the SAS, I had the privilege to meet a Colonel in Northern Ireland named Anthony Hunter-Choat. As a young man, Tony had left school and decided to tour around Europe, finally ending up in Paris, where he joined the French Foreign Legion. After training in Algeria and becoming a parachutist, he was duly posted to the 1st Battalion, Régiment Etranger de Parachutistes (1e REP), with which he would be involved in continuous operations for almost five years. Wounded and decorated many times during his early career, he was cited as one of the plotters when the army tried to seize control in 1961 against de Gaulle. Subsequently 1e REP was disbanded; as its men were marched out of camp they sang Edith Piaf's "Non, Je Ne Regrette Rien." Shortly afterward Hunter-Choat's five-year term of service expired and he returned home.

In March of 1962, despite being overaged he managed to join the British Army and get commissioned into the 7th Gurkha Rifles. A year later, he was in Borneo in what was known as the Indonesian Confrontation, where small patrols would carry out cross-border operations. Then, after a small spell as a battery commander in 3 Royal Horse Artillery, he was offered the command of 23 SAS Regiment.

It was about this time that I met Tony in Northern Ireland, where he was working for one of the security services. He was a soldier's soldier: extremely likeable and a hard worker with a gift for motivating his men. After a brief spell at NATO head-

quarters, he retired from the British Army, and went to work for the Sultan of Oman. In 1995, Tony was presented by the Sultan with the Omani Order of Achievement. After the American-led invasion of Iraq in 2003, Hunter-Choat became head of security for the Program Management Office (PMO), which was involved in overseeing the distribution of billions of dollars of reconstruction funds to projects throughout the country. There he briefly became embroiled in

Tony Hunter-Choat, a man who knew what he wanted from life and went out and got it. He was one of the best and most intelligent soldiers I have ever had the privilege to meet.

controversy after the PMO awarded a contract worth $293 million to Aegis, a private security company headed by Tim Spicer. According to *Vanity Fair*, Hunter-Choat and Spicer had known each other for years. DynCorp, a rival to Aegis, lodged a protest with the US Congress, but this was rejected as there was no suggestion that Tony had behaved improperly. He was later responsible for the security plans for US Aid in Afghanistan. Sadly Tony Hunter-Choat died on April 16, 2012, but was a man who had lived his life to the fullest.

Work for a Private Security Company

Working for a private security company is a similar method of disappearing as joining the French Foreign Legion, but without your family fretting too much about you leaving. Other than the fact that it's a dangerous life, there is no real downside, and the pay is excellent.

Working for a Private Security Company (PSC) is an exciting life and extremely well paid, but you will need the right skills.

In order to become employed by a Private Security Company (PSC), you either need a military background or the skills that they need. It is possible to learn the skills required, but it is difficult. The best advice I can offer can be found in one of my recent books, *Soldier of Fortune Guide to How to Become a Mercenary*. It's not really about mercenaries, but more about the modern day private security officer. Not exactly the best way to disappear, but you will find yourself in places where few people will find you.

Join a Religious Order

There are those people out there who would happily swap their current condition and life for the peace of a religious order. There are many religious orders around the world, and if you are a true believer, then you should try this for a few years. Provided that you enter this with your eyes and heart wide open, there are no drawbacks to this type of disappearance.

It is possible to join a religious order in many countries, and few ask more than you are sincere and honest. For example, a person

The Taktsang Palphug Monastery (also known as The Tiger's Nest), is a prominent Himalayan Buddhist sacred site and temple complex, located on the cliff-side of the upper Paro valley, Bhuta. You could always find a hideaway like this, but unless you're really devout, life would be boring.

who wants to become a monk must dedicate their life completely to the service of God and must give up just everything they own to live a monastic life in a place of prayer. There are many types of orders in various religions, such as Catholicism and Buddhism.

The path to becoming a monk can take many years of commitment and study, so it is not something that one would enter lightly. In addition, you will have to dedicate yourself completely to the chosen religion. It is, however, a good way of disappearing completely—albeit a little unorthodox.

To become a Catholic monk, a person must be of honorable intent and sound of mind and body. He must be a Roman Catholic and have received the Sacrament of Confirmation; a ceremony usually performed in the adolescent years that makes a person an official part of the Catholic Church. Before joining, you must also make sure that you are leaving behind no responsibilities that can hound you, such as heavy debts or a criminal record.

The first step to becoming a monk in the Catholic Church is to visit a monastery, preferably several times. These visits should help the person decide if this is a life they could happily lead. The next step is to contact the Novice Master, who is responsible for overseeing the training of new monks. For some people, becoming a monk may be the ideal way of disappearing, or at least of getting yourself into seclusion where you will be safe and can live the rest of your life in peaceful tranquility.

It is also possible to travel to a foreign country and become a monk or nun. In this role, you would spend most of your life in prayer and being of benefit to others. While most Buddhist monks are in Asia, there is no barrier to a Westerner who believes in Buddha from becoming a monk. Although there is a great benefit in the life of a monk or nun, you must bear in mind that in doing so you are taking on a big and deep responsibility for yourself and for others.

No matter which religion you choose, you should always do your research before making any decisions, such as the various criteria for entering a specific religious order. Traditionally, a student requesting ordination will have completed several years of study and practice under the guidance of a qualified teacher. If your plan to disappear involves joining a religious order, then you should take your time to develop and understand the teachings of the religion you are about to join.

In the case of a becoming a Buddhist monk, it is possible to spend some time living in a monastic community where you will receive religious advice from resident monks or nuns on what it is like to become monastic. Always keep in mind that becoming a monk—even though your end in all this to disappear—is a serious undertaking. The vows of a Buddhist monk or nun are taken for life; therefore it is important that you reflect seriously before you go down this path.

For many years, I have been dealing with a company in Bangkok that specializes in military equipment. One of the salespeople in this company surprised me one day by revealing that he had been a Buddhist monk. He spent many years in India and Thailand solely dedicated to the Buddhist order. Then one day he simply walked out of the monastery in nothing but the clothes he wore with nowhere to go, no money, and he was all alone. Through a translation job, he became employed by the military company, and to this day he is one of the happiest people I've ever met. He is still a dedicated Buddhist, but now is married and has a family with a whole new life ahead of him.

You could always move to a foreign country and live in the wilderness. Your chances of being discovered are slim.

Move to a Foreign Country

Moving to a foreign country is the number one method for disappearing. However, it takes a lot of planning, preparation, and determination, which few people accomplish. If you do it right, there are no real drawbacks and a new life awaits you. I will keep this section short, as the rest of this book revolves around most things you will need to know if you decide to take this route to disappear. My only comment is to plan very carefully where you want to end up, and do a lot of research beforehand.

Places to Go and Places to Avoid

It does not matter if you opt to move to a foreign country, join the French Foreign Legion, become a mercenary, or join an NGO: at some stage you will end up overseas. The secret here is not to end up in a place you really don't want to be. Trust me, there are some great places in the world, but equally there are some dangerous crap holes—Northern Nigeria being one and Syria being another. Should you find yourself in places like this, you are in serious trouble.

Author's Note: As I write this, the British news has announced that seven Christians who were working for a construction company in Jama'are have been kidnapped by Islamic terrorists and executed. This is not an isolated incident, as the Ansaru (a splinter group of the main Northern Nigerian terrorist group Boko Haram) had been raiding and slaughtering any foreigner they could find in the area.

These are the poorest countries in the world, and places you should avoid. Haiti has over 77 percent of its 10 million person population living in poverty. There is little or no work, disease and pestilence are everywhere, and it has a crime rate that includes rape and murder on an hourly basis. Other countries to avoid include Equatorial Guinea, Zimbabwe, Congo, and Swaziland; the latter has a very low survival rate and the average life expectancy is just forty-eight years old.

You should also do a background check on any country you intend to visit just to ascertain how many people involuntarily disappear each year. While it should not affect you, in some countries people fall foul of government and thus are forced into disappearing, i.e. they are killed and their bodies disposed of. Organizations such as Human Rights Watch and Amnesty International do their best to keep track of any forced disappearances and frequently publish their reports.

Chile, under the dictatorship of Augusto Pinochet, saw the disappearance of literally thousands of people—both from the political opposition as well as anyone else he did not like. Many were tortured before being killed in what became known as the Caravan of Death. In truth, this was a group of army officers who flew the length and breadth of Chile by helicopter. They would stop at prisons or army garrisons where anti–government activists were held; each time they would select their victims who were then murdered in cold blood before being disposed of in an unmarked grave.

By comparison, there are many great places to start a new life. You may notice I mention places such as Canada, Malaysia, and Thailand a lot in this book. Apart from my personal knowledge of these places, they are, in fact, brilliant countries to disappear to. Ideally, you need a country that's friendly to foreigners, has great weather, and an easy lifestyle with little or no pressure from the authorities (unless you have done something wrong, that is).

Venezuela is a really good place to disappear, and even if you get found by some skip trace expert, the chances of you being extradited at the moment are very slim. There are some 30 million people living in Venezuela and many come from Europe, Africa, and a dozen or so other places around the world. Because of this, it makes no difference if you're black or white, as you will not stand out. In addition, the recently departed Prime Minister Hugo Chavez would not allow any Venezuelan to be extradited (unless it suited him) so if you can gain citizenship after five years, you will be untouchable even if you are found (that could change now that he is dead).

Other than a good ethnic mix, Venezuela offers lots of other advantages: the country runs mainly on hard cash, so financial records are few and far between. Likewise, you can put your card into an ATM machine without much chance of a trace being set off. The weather is just about perfect and the country offers mountains, desert, sand dunes, and sunny beaches for your pleasure. For those of you who are single or thinking of leaving a partner behind when you disappear, Venezuela has lots of healthy male and female companions waiting. In fact, it's hard to walk down a street in Caracas without seeing a beautiful woman who will give you a wonderful smile.

For those of you who think you're likely to get kidnapped the moment you step off the plane or killed on a street fight between rival drug gangs, forget it. While it does happen, you will find the people in Venezuela to be some of the friendliest in the world.

You will need to do some research before deciding to which country you will travel. Some are very poor and the lifestyle may not be to your liking.

I spent a wonderful four months in Venezuela and enjoyed every minute of my time there. It is certainly one of the places I would opt to disappear to. However, it would prove beneficial to have at least $50,000 at your disposal, as this will buy you a little time to get used to the place. Be frugal with your money; don't flash it around, and don't carry it all with you. If you should find yourself in Venezuela, go and see Angel Falls. It's really worth the visit, as it is the highest waterfall in the world.

Your Safety in a Foreign Country

Feeling safe and secure is very important. A backstreet in Beirut can be a peaceful marketplace during the hours of daylight, with the bustling streets offering a degree of protection and normality. However, at 2:00 a.m. the market traders will have gone and the street will be empty.

If you find yourself in a semi-hostile country, stay in a safe hotel and don't walk the streets at night. This is my hotel in Tripoli just after the revolution.

When you arrive in a foreign country, the first question you must ask yourself is, "Why am I here?" Assuming you started in a safe location, what brought you to an unfriendly one? No one simply walks into danger, but the activities of someone disappearing might require you to do so. In planning your disappearance, you must understand the dangers both known and assumed, and make preparations for your safety. Should you walk, drive, or use public transport, and what is your alternative if things get rough?

Author's Note: A couple of months ago I was in Libya. The streets were full of young armed men sitting behind heavy machine guns that had been fitted to Toyota pick-ups. This militia controlled the country, and it seemed that no individual was in charge. Feeling as though this was not a good situation, I stayed in my hotel and took a taxi everywhere. The area was known to be hostile (although being British, whom along with the Americans were well liked), and at night the side streets were full of small armed gangs. Normal activities calmed down, i.e. the local population moved about their business. Still, when you are the only stranger in the immediate vicinity and all eyes are on you, what do you do?

The logic of both geography and time provide us with situation awareness, and this is what you must learn and react to if you are to survive. Situation awareness is a mixture of visual and metal simulation triggers.

1. Ideally, at the first signs of a troublesome situation, you should walk or drive casually to the last known safe area and extract yourself. (In my case, the British Embassy, with whom I had already registered my presence.)
2. If this is not possible or your way is blocked, you must look for an escape route.

3. If none are available, prepare for an imminent attack, but keep moving. Always head for larger buildings as these indicate either government presence or major shopping areas where there are more people around.
4. Aggressively confront those blocking your path, but don't start the confrontation.
5. If you have to fight and flee, do not become subject to capture. (Take aggressive actions.)

Don't worry too much about them all having weapons. Trust me, most will not have them ready for instant use, and you only need to put a few meters between you and your assailants for them to miss if they do shoot. Few people are accurate with a short–barreled weapon at a range of more than 10 meters. That said, it has been my experience that if you do run, they will chase you because running is an act of guilt. If you walk away quickly, they just may not follow you and let you be. In such a situation, you are the only one who can make the correct decision.

Summary

It does not really matter how or where you go just as long as you keep in mind the end goal: to disappear and never be found. Becoming a monk or a hobo, driving around in your RV, or any of the above will really be down to you. Be innovative, think of a way to disappear and start a new life style, do your research, and look at how others have managed to do it.

Give yourself a trial run—go for a three month walkabout, plan to raise money for charity, or walk the entire length of America. Some years ago, this would have been a well-publicized event and open to exploitation and broadcast by the media but not any longer—everyone is doing it. See what reaction you get from your friends and family if you choose to disappear on a fund–raising walk

One simple way to disappear is to have a quest. Mike Howitt set himself a goal of walking to South Africa.

for three months, as this will give you some idea of what to expect when you really disappear for good.

Mike Howitt, a pensioner of eighty years old, walked more than 10,000 miles to South Africa to scatter his wife's ashes in the country she loved. Mike undertook this task as a labor of love and to fulfill his wife's wishes. During this 10,000 mile walk, he passed through twenty-one countries, starting off in Leicester, England, and finally he made his way to Cape Town. He suffered many hardships and at one point spent twenty-two hours in a cattle truck, but never at any time was he threatened by violence or disaster. The total journey took Mike just two months to complete, during which time he raised over £6,000 for a local hospice.

7

HOW YOU GET CAUGHT

You will get caught for one of two reasons: because people are looking for you or you do something that gets the media's attention. If people are looking for you, there is a good chance they will find you. Professional skip tracers or private detectives have a whole arsenal of methods that will help track you down; if it's a government agency that's after you, then there's no chance of hiding.

Staying below the radar is hard and requires discipline. Even with the best cover in the world, we can all make mistakes or do something that brings us to the attention of the media. A simple accident where you crash your car into a motorbike and kill someone might only make the local newspapers, but the fact that you are an American from Oregon might just make *The Oregonian*. Trust me, this happens quite often.

Keeping a low profile is extremely important. I was actually in Bangkok when Viktor Bout was discovered and arrested.

If you get involved with any shady characters, crime, or drugs and get caught by the local police, it's a strong bet that once they discover your identity and nationality, they will inform your native Embassy that you are under lock and key, at which point you are back in the system. Then the fact that you've disappeared for five years will become known to all.

In order to avoid these two main forms of compromise, you will need self-restraint and a change of habit. We are all creatures of habit, and it's very difficult to change our ways, but if you intend to disappear for good you will have to make a serious attempt to change your routines. Let's take a look at what most people will do unless they steal themselves and do the unexpected.

- If we tell anyone of our plans, it is likely to be our mother, father, granny, brother, sister, and so on. Why? Because they are close to us, and we would expect them to protect us. They will eventually talk.
- If we go anywhere, we always tend to go home. If we go home, we will frequent our old haunts and bars. We will immediately be seen and recognized.
- If we earned our living in the old life as a car mechanic, then we will more than likely adopt the same way of making a living in our new life. That's an easy way for someone who knows your history to find you.
- Consider what will happen if you become involved in crime in your new country or get put in jail. It's no good pretending you are a local—Americans and the British stick out like a white ball on a pool table.
- Negate any risk of accidents. In other words, don't have a car or motorbike (a local scooter is an acceptable risk). Use public transportation to travel.
- Don't get married to a local, as this could eventually require your presence at your local embassy.

Embassies

The United States, as well as many other countries, maintains embassy offices in countries all over the world. The office serves an important function in friendly foreign relations between the home and host governments. One of the functions of any overseas embassy is to protect its country's people. The embassy can be a point of contact or base of communication between two countries. Embassies are also keen on knowing exactly who is in the host country, i.e. you. While they may not know of your presence, others who are in contact with the embassy might. That is to say, one American might meet another in a small village in Bolivia quite by accident. The

Thai girls are among the most stunning women in the world, but try taking her home and you will need to make your presence known to the embassy.

stranger might well forget you the moment you part, but they may also mention it when visiting the local embassy.

Always be very wary of your local embassy. While they are there to help and protect you under normal circumstances, if you want to disappear for good, stay off their records. Side-step accidents that may bring them to investigate you and avoid getting yourself arrested. Finally, never try to marry a local girl; should you split up for any reason you can bet your bottom dollar she and her family will make the embassy their first stop in order to locate you.

When the Government Wants You

Why would the government come after you? If you have done nothing illegal, committed no crime, or have not tried to avoid the law, then there is no reason why the government should try to locate you. That

They are the people you never see and rarely hear of, but believe me they are there and they are watching. This is GCHQ in Cheltenham—the watchful eye of the British end of Echelon.

is not to say they will not, but the odds are they won't use their valuable resources to trace someone with whom they have no interest. However, if you are of interest to the government, especially in America, Great Britain, Canada, Australia, or New Zealand, you could fall foul of some serious tracking. While each government has a vast array of methods for tracking people down, there is also a huge computer network running 24/7 that can find you called Echelon.

Echelon

Echelon is the name given to the massive worldwide surveillance system that is capable of capturing and scanning every telephone call, fax, and email sent anywhere in the world. Using sophisticated satellite systems, earth stations, radar, and communication networks, as well as an array of ships and planes, the system is capable of monitoring both military and civilian communications. It was originally developed during the Cold War by English–speaking countries to eavesdrop on the communications between the Soviet Union and its allies. As that need is no longer pressing, it is instead being used to monitor terrorist communications, as well as the activities of organized crime alongside the usual espionage territories of political and diplomatic intelligence.

Although details about the system are still shrouded in secrecy, some facts are known. The main proponents are the US and the UK, but they are backed up by Canada, Australia, and New Zealand. Each country is responsible for monitoring a certain part of the Earth. For example, the US listens in over most of Latin

America, Asia, Asiatic Russia, and Northern China. Britain monitors Europe, Africa, and Russia west of the Urals. Canada sweeps the northern parts of the former USSR and the Arctic regions. Australia is responsible for Indochina, Indonesia, and Southern China, whereas New Zealand handles the Western Pacific.

The way Echelon works is simple in practice. All members of the alliance use satellites, ground receiving stations, and electronic intercepts that enable them to pick up all communications traffic sent by satellite, cellular, microwave, and fiber-optic means. The communications so captured by these methods are then sent to a series of supercomputers that are programmed to recognize predetermined phrases, addresses, words, or known voice patterns. Anything deemed to be of interest is then sent to the relevant intelligence agency for analysis.

In the US, the agency responsible for Echelon is the National Security Agency (NSA), based at Fort Meade, near Washington, D.C. It is estimated that it has a staff and resources in excess of the combined CIA and FBI budgets. Canada's Echelon program is handled by the Communications Security Establishment, an offshoot of the National Security Agency, and is based in Ottawa. In Britain, Government Communications Headquarters (GCHQ), located at Cheltenham, is concerned with Echelon. But it must be borne in mind that the locations of smaller stations are spread across the globe in strategic positions.

After the Cold War and before 9/11, the main thrust of usage for the US was to intercept messages from South and Central America in an effort to thwart drug barons from exporting their lethal and immoral cargoes. Other organized crime gangs and terrorists, such as the Russian Mafia and Hamas, were also a target. However, post–9/11 it must now be assumed that Echelon is on alert for any messages that might warn of an attack by Al Qaeda. Although such usage of a surveillance system can only be a positive thing, it has also had its fair share of detractors. Certain accusations have been made that Echelon has been used to commercially benefit the countries involved, enabling them to undercut competitors and double deal to national economic advantage. Debates have even been raised in nonparticipating countries and within the EU. Nevertheless, the intelligence gains provided by the system in the new climate of global terrorism are likely to drown out any protests in the future.

Private Investigators and Skip Tracers

Even if the government has no interest in you, your family and friends will and may hire a private investigator or a skip tracer to locate you.

Skip tracers are private investigators who act mainly for finance companies. Characteristically, they search for people who owe money and have subsequently done a runner to try to avoid paying it back. They are very good at what they do and know just about every trick in the book with which to find you. The finance companies only hire the best, so if you have one on your tail, you will need to have stuck to your plan to not get discovered.

A good skip tracer picks up the phone, trawls the Internet, talks to your family, friends, and neighbors, and establishes a profile on you. Whether they actually find you or not will be down to the success of your plan to disappear, as well as what clues you forgot to cover before you left. Remember, information is only as good as that provided and attainable by the skip tracer agent.

Find a very close associate of someone who needs to be traced—mother, father, wife, child, girlfriend, brother, or sister—and the odds are you

Barry Davies

It is my belief that we keep more information about our personal lives around our office desk than at home. This is always a great place to find out information on someone or about something.

will find information that will lead you to your target. Unless you have built your disappearing plan correctly and stick to the golden rules, you will be found.

Golden Rules

The golden rules are the basic fundamentals that you must obey if you want to disappear completely and never be found. In the military there are a set of rules called Standard Operating Procedures (SOPs). SOPs are built on previous operational experience and are the backbone of any successful military operation; in the SAS, you break these SOP rules at your peril. If anyone finds out—and they will—you will be thrown out of the unit.

It is the same for anyone thinking of disappearing: follow the basic rules and your plan will work. Your prime and only goal should be to terminate the old you without any trace of suspicion, and reemerge as the new you where you can start afresh without ever being discovered. In order to do this, you must follow the golden rules:

- Be committed that disappearing is the right thing for you to do.
- Build and test your disappearance plan.
- Sever all previous contacts—leave no trace behind.
- Unless it's part of your plan, give no hint of your impending disappearance.
- Build a new identity and live that identity.
- Avoid accidents and criminal activity.
- Stay away from foreign nationals, i.e. Americans, British, Europeans.
- Stay away from your country's embassy (unless you're in serious trouble).

Believe me, all of this is easier said than done. Once again, it's down to the commitment of the individual.

Author's Note: I was once in a very short-term relationship and made the terrible mistake of marrying someone I shouldn't have. While the courtship and the marriage lasted only a year or so, I saw the error of my ways within the first six months. One night as I sat and listened to the verbal abuse pouring from her drunken mouth, I decided to make a clean break—but on my terms. I rented a small office on the pretext that I could work better, and this was an acceptable solution while she went and socialized. I moved everything that was important to me into the office: my desk, computer, and books. Slowly I managed to retrieve all the items from my home into my office. Then one day I simply called her and told her it was over. It might sound like a coward's way out, but in a face-to-face confrontation, she would have tried to force me to be violent (some women want violence so they can go to the police and get you arrested). Also, she had two great kids from a previous marriage, and confrontation in front of them would not be good.

We had no joint bank accounts, and she had plenty of money of her own—plus she still had a home and a car. So I simply walked away and went to Spain. True, she could have found me if she really wanted to, but she knew it was over anyway.

Becoming Noticed

Talking to expats in a foreign bar is a quick way to get yourself noticed and impart information you should not.

It is easy enough to undo all your good work. One of the biggest mistakes you can make is to become friends with like-minded foreign nationals. While it's perfectly normal for one American to strike up a friendship with another in a foreign country, it can ultimately lead to your discovery. Sure, it's great to have company and conversation, but over a few drinks a lot can be said. One of the first questions a stranger will ask is "What are you doing in this place?" No matter how careful you are, you are already putting cracks in your disappearanc plan. Even if you avoid the truth and tell a plausible lie, you still open yourself up for more questions. The conversation will lead to "Where do you come from in the States?" And so the fabric of your plan starts to crumble. Remember discipline: once you are below the radar, stay there.

Generally, there are good reasons behind the disappearance of thousands of people every year. Their state of mind, unbearable pressure, financial worries, love affairs—the list is endless. But when you look at each case individually, there are pointers that will identify why the person disappeared. One example is people who disappear from a cruise liner: you either get off the ship voluntarily, are thrown off by a third party, or simply fall overboard while under the influence of alcohol or drugs. Likewise, you may be under threat; many a worker for an NGO has disappeared while working in Africa—it's a dangerous place. You may work against the country's authority, and your removal could be the answer the ruling government requires. Trust me; it happens more than you think. Below are just four cases of people who have disappeared, and it's easy to identify a plausible reason in each instance.

> Richey Edwards, a twenty-seven-year-old member of the Welsh Rock band Manic Street Preachers, had a history of self-harm and disappeared in November of 2008. For years he had received treatment for depression and alcohol abuse, so it came as no surprise when his car was found abandoned close to the Severn Bridge, which connects Wales to the UK. He had jumped from the bridge and was presumed dead.
>
> A Swiss–born activist, Bruno Manser, who fervently campaigned for the preservation of the rainforest in Sarawak, disappeared after leaving the small isolated village of Bareo in the Malaysian state. Bareo is very close to the

Indonesian border and, having spent some time in this desolate jungle where the only occupants are descended from ancient tribes of head-hunters or patrols by hardened military forces, it is simple to see how someone could

Bruno Manser in the same jungle where he was found dead.

disappear. It may have been a simple accident where he fell and broke a leg, or he might have been savaged by an animal; we will never know. He was declared legally dead in March of 2005.

The Thai Muslim lawyer and human rights activist Somchai Neelapaijit represented many of the insurgent terrorist groups operating in Southern Thailand. He simply disappeared when he visited Bangkok on March 12, 2004. Given the state of anti-Muslim feeling in the capital and his support for the various terrorist atrocities that had taken place, there is reason to believe that his disappearance had been a planned operation.

Closer to home, the 20th Century Fox executive Gavin Smith disappeared in May of 2012. It is reported that he left a friend's house in Oak Park, California, and was last seen sitting in a restaurant a week later with an unidentified woman. This fact throws doubt upon Smith's disappearance, and it is thought that he is still alive.

Stay Below the Radar

How easy is it to find people? Well if you're not trying to hide, it's extremely easy. As I sit here writing this book, it came to mind that I've never seen my editor, Jason Katzman, at Skyhorse Publishing.

So I simply searched for his name in a browser on my computer and was presented with a list of options, from which I selected Jason Katzman, Assistant Editor at Skyhorse Publishing. Now I know what Jason looks like, where he works, where he was educated, and can see most of his friends and work colleagues. That is a lot of information just at my fingertips, which took seconds to find.

You can do everything within your power to remove traces of your current life from your home and possibly your office. However, it is almost impossible to remove items that show your current face and information from the Internet or held on file by a myriad of government and civilian agencies and companies.

Mobile Phone

Your mobile phone is another source of information for anyone who wants to know where you are. Should I wish to steal or check on someone's identity, this is the first thing I would get my hands on. Most of us keep so much information on our phones that it's almost better than gaining access to someone's computer. Modern smart phones will not only keep you up to date with your social network, but you can also do your banking online. These two services will supply anyone with enough information about you and your family to enable them to steal your identity or find out nearly everything about you. True, there are a few password loops to jump through, but these are easy to bypass. It will depend on the system you are using, but if you have a look at the skills I mentioned in this book, you will see how to bypass a password on a mobile phone and computer.

Getting ahold of anyone's mobile phone is very easy. Let's face it, we all carry it with us, leave it on our work desk, sit in a café with the phone by the side of our coffee. The Lookout phone app reported around nine million lost smart phones in 2012, with around one million being lost or stolen in the UK. On average, most of us lose or have our phones stolen at least once a year.

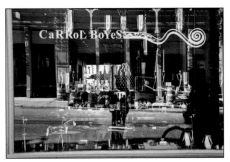

If you think you are being followed, use a shop window reflection to see if there is someone behind you.

If you want to stay undetected once you have disappeared, but you really must have a mobile phone, my advice is to buy a simple mobile, not a smart phone. Memorize your contacts—there should only be a few—or if necessary keep them on a separate list in a safe place. Remove or delete any information on your phone that might lead to your real identity.

Being Followed

One of the first skills you must learn after your disappearance is that of good observation. No matter where you are or how safe you feel, you should be systematically recording places, events, and people so that you'll be able to recall them with accuracy. The military have a lesson on this subject. They place a number of objects out in an area some meters away from the trainees. They are given a short period of time in which to observe all the objects before they are taken away. The trainees are not allowed to write them down for several hours. Later in the day, they are asked to name all the objects and the unique features of any item.

If you are being followed by professionals, there is little chance of escape. Your best and only course of action is to throw them off your trail and then put as much distance between you and them as possible. Government surveillance teams may well have access to your original fingerprint records or your DNA; sitting in a bar having a quiet drink will leave your DNA on the glass, and that's all they need. Countersurveillance is an art in itself and far too detailed for this book, but if you should get suspicious, it's always best to take precautions rather than do nothing. You can train yourself to be very observant, and it can be highly entertaining.

Try doing memory exercises, such as when you enter a room for the first time, by remembering a car number plate, or noting the number of windows in the house you just passed. The trick is to keep as much information in your brain for as long as possible. As with most subjects, the more you practice, the better you become. Most importantly, start to recognize faces in a crowd and note people you see in the same place at the same time every day. You will be surprised at how much your brain can absorb.

No matter how confident you are about covering your tracks, if professionals are trying to find you there is always a chance that they will. Make sure to adjust your attitude and appearance to the situation and think before you react. What do people expect of you, what role model have you adopted? It's no good pretending to be a businessman if you're dressed like a tramp. You must play the part you have planned for in order to avoid suspicion. If someone is looking for you, they are looking for an individual—try to blend in with your surroundings and the people.

Be interested in people, as this takes the focus off of you. Most people are only too happy to boast about their social position, wealth, family, or occupation; take advantage of this. Learn to listen to your sixth sense or analyze any gut feeling you might get. There are many basic instincts in the human brain that warn us of danger; learn to recognize them and take appropriate action. You may walk down a street and recognize the same stranger you saw only an hour ago in a different part of the city—is this coincidence? Walk down the street, then stop and look in a shop window. This way you can use the reflection to see who is behind you. Is the stranger still there? If the answer is yes, then walk on normally, DON'T run, turn the corner, and distance yourself as quickly as possible.

If you think you are being followed, my advice is to lay low for a few days, stay in your accommodation and not venture out, or simply catch a train and get off at a stop some miles away.

Only ever take calculated risks—never be a gambler. With a calculated risk, you can spot the drawbacks and adjust your plans accordingly. Always analyze your actions and base them on solid information. If you take a gamble, you only need to fail once. Here are a few tips:

Catching a train or bus is a good way to avoid being followed, but don't rush. Do it naturally, that way those following will not get suspicious.

- In a confrontational situation be aggressive, as no doubt the skip tracer will be. Let them know you are not to be messed with, and get your punch in first.
- If you have to speak to or shout at anyone, do so in the native tongue. This might confuse the skip trace agent who may think he or she has mistaken you for a lookalike.
- Know your own strengths and weaknesses.
- Know your territory and its inhabitants; use them against the skip tracer.
- It is better to be known than be a stranger to the area and its inhabitants. Let your cover story protect you. Make good local friends.
- Know when to get out, and always have an escape route planned.
- When the situation goes "pear shaped" and you get caught, have a backup plan.

The Cover Story

Before disappearing, you need to make a cover story for yourself. The best way you can learn to do this is to act like a spy would. In the world of espionage, the one thing that must stand up to scrutiny is the spy's cover story. They must be who they say they are and, when working in a foreign country, be able to prove their identity. By far the best way of obtaining a cover story is to make it as near to the truth as possible. For example, details such as your age and place of origin, your education, and your likes and dislikes. By doing so, you do not fall into a trap when a skip trace agent starts an in-depth background check of your life.

Many spies enter a foreign country as part of the embassy staff or as a member of a diplomatic mission. In some countries, the role of the Defense Attaché is just short of a spy master. His position will not allow him to partake in direct actions, but he can act as an umbrella for a network of spies and agents working on behalf of his country. From time to time, spies are recruited by a government simply because they have the right qualifications. They could, for example, be a business-man who has just won an order with a foreign country. This grants him an automatic cover story and a legitimate reason for traveling. However, both of the above examples are restricted, firstly by protocol and secondly by lack of espionage training. The real answer is to train a potential spy in the arts of tradecraft and provide him or her with a believable cover story.

Barry Davies

This is Captain Robert Nairac. I was with Bob the day before he went undercover into a pub frequented by the IRA. They caught him and after some horrendous torture, killed him. His body was never recovered. A captured IRA man some years later said Bob was one of the bravest men he had ever seen.

Example Cover Story

An English spy was sent to work in the border country of South Armagh in Northern Ireland. His accent was English, as was his manner, but he managed to operate and collect information from the local farmers for six months. How did he manage this?

He adopted the role of a salesman, selling impactors for an American company, which had an overseas office in Belfast. The impactor, which fit the back of most standard farm tractors, was designed to break up old concrete. It was a solid cover story; he even arranged for a demonstration of the machine at a local agricultural show, which added credence to his presence in a known terrorist area. His explanation was simple: the company had afforded him six months in which to establish the impactor, after which his job would be in jeopardy. For their part, the American company was keen to have someone try to sell their implement, for everyone knew it was far too expensive.

The spy did his homework, first by obtaining good road maps and aerial photographs of the area. Second he researched how and where the IRA had been active; he also did the same for the local British armed forces. By doing this, he could establish a route into the area he wished to visit without being stopped either by the IRA or the British army. The latter would not pose a major problem, but in the eyes of the locals it was best to stay clear of any association. His visits were all done in daylight hours, so as to avoid being stopped by the IRA in an evening roadblock.

Although his movements around the area looked casual and random, they were meticulously planned. He managed to visit most of the farms and smallholdings in the area, taking in a local pub at lunchtime. At first, his reception by the locals was mixed; some accepted him immediately while others eyed him with suspicion. The most common question that arose was "What are you doing here? You know this is a dangerous area." This was a perfect question, which allowed the spy to open up the conversation in regard to the IRA. He would respond, "That's all newspaper stuff; I have not seen any problems." Then, the farmer would fall into a conversation recalling all the deeds of the local IRA. If the conversation went on for more than five minutes, the spy would casually ask the farmer if he would like a drink, saying, "I have a bottle in the car." Few refused.

Within two months, he had built up a list of friendly farmers and isolated his lunchtime drinking to one particular public house. Having assessed all the people he had met, the spy set about honing in on several, the first being the daughter of the publican. She was a woman of about twenty-four years old, good looking, and full-bodied Since her father had died when she was very young, she helped her mother run the bar. Friday and Saturday nights saw her receiving a lot of attention from the local young men, who attended the pub from both sides of the border. Because of this, she had not ventured far from home and had seen little of the world. As far as she was concerned, the spy, who was thirty-five years old, had appeared in the pub like a breath of fresh air. The spy had noticed the attention whenever he entered the pub, but in the beginning he played it friendly and low key.

One lunchtime, the spy discovered early in the conversation that her mother was away for several days. The spy turned on the charm and the woman fell for it. Closing time was 3:00 p.m., but she indicated that he could stay if he wanted—he did. In the four hours until the bar opened for the evening, the spy made love to her three times—she was hooked. At 7:00 p.m. the bar opened; the spy had one pint and left. As he drove back to his safe house in Armagh City, he mentally recalled all the names the woman had mentioned.

The liaison endured secretly for three months and each time they were alone the spy would ask his seemingly casual questions, all of which had been carefully rehearsed. A miniature microphone faithfully recorded every word the woman said. The information she had given was predominantly about the young men she had known since they became of drinking age. They frequently used the pub and had tried to impress her with their stories of heroism by pretending to be members of the IRA. For the most part this was just bravado, but she knew that one or two of them spoke the truth. "Be careful of him when he comes in," she would tell the spy. "He's a real nasty piece of work."

One day, the spy simply never came back, and after a time he was forgotten. He had managed to infiltrate a dangerous area by using a substantial and plausible cover story. The spy had taken time to get to know the area and the inhabitants before asking any questions. With the use of alcohol he had gained information from the farmers, and with charm he had gleaned valuable information from the pub owner's daughter. Despite speaking with an English accent, his cover story held together and was believed.

Tell people at work about your plans to fulfill your desire; boating, a survival course, learning a language, and so on, but never give a hint that you intend to disappear.

Strengthening your Cover Story

One excellent way to strengthen your cover story is to find yourself a new partner. This is one of the most important aspects of forming a new life for yourself, as it will help to establish your residency in a country. Being married to a local will also assist in covering your tracks. Finding and living with a new partner will help with your accommodation, local knowledge, local society, and establishing a new life. All of this will hinder anyone trying to find you.

Depending on your age, looks, and health, there are a number of ways to find a partner, but in most civilized countries the easiest way is the Internet. Online dating can be found in just about every country, including some of the more strictly religious ones. The amount of really nice people out there who simply want a partner to love and care for them runs into the billions (see Chapter Ten).

Author's Note: David is a friend of mine who was divorced and looking for a new direction in life. He found it when the company he worked for offered to send him to Malaysia, where he would set up a new office for them. David moved out to Kuala Lumpur and set up his office in the center of the city; he also rented a four bedroom house in a nice residential area. While the night life in Kuala Lumpur is excellent, David eventually found himself a little lonely living alone. It was during a discussion with one of his Malaysian customers that he learned of a good dating website and decided to try it out. At this time, David was around thirty-eight years old, with a good build, fresh faced, and relatively handsome. In addition, he drove a new BMW and lived in a superb home. All this information went on the dating website.

David married a Singapore Airlines stewardess (it's still the best airline in the world).

When I met David some six months later, he was sitting in his house with the most stunning girl snuggled up to him. He told me the story of how they met. Within a matter of minutes of him uploading his details, he had some five hundred hits. By the next morning, this had gone into the thousands. Smugly, he went through the offers and selected the cutest girls living within the city limits. The next few weeks were a bit heady as David went out almost every night with a different girl, all of whom wanted to be his new partner.

Then he met the woman sitting next to him on the couch. She was a stewardess with Singapore Airlines and just about as gorgeous as a woman can be. They married. David even converted to Islam to placate her parents (it was in name only, as he still drinks), and they both live happily to this day.

Summary

Once you have disappeared, stay below the radar and remain anonymous. Don't do anything that will bring you to the attention of the authorities. Guard all your information and don't leave useful intel on your laptop or mobile phone that others can use to identify who you really are. If you think you are being followed, check it out without them knowing you have spotted them. If you act like a fugitive, then you are a fugitive. That said, a few tricks to instantly disguise yourself will always help throw anyone following you off the trail.

Embrace your newfound life and befriend the local population, speak their language, marry a local, learn, and become part of their society. Most importantly, stick to your cover story. If you have planned it right, there is no reason why people should not believe you unless they can do some serious background checks, such as DNA testing.

8

DISAPPEAR AS A HOBO

If you are hell–bent on disappearing and cannot wait another year to plan a proper disappearance, then become a hobo. This is the way I would do it: Just walk away to a place where there are no people, no ATMs, and no problems other than surviving. Live on your own initiative; do your own thing. Remember that becoming a hobo is just a means of disappearing; it does not have to be forever, as you can always return a few years later. You don't have to change your name, get a false passport, or plan any trips overseas. Should you wish to return, you can always claim to have had amnesia!

You simply leave one day and never come back. Your disappearance would be more convincing if you laid down a simple deception plan and took a little time to organize your disappearance, but if you really wanted, to you could simply go now. Move away as fast as possible; put as much distance between you and your home location as you can. If you have time, do a little planning, as this will help. For example, you could withdraw some money and drive an old car north to the Canadian border. This 3,987–mile border is the longest undefended border in the world. While there are border guards and drones flying around, if you pick your spot in the forest and cross around 5:00 a.m., you will have no trouble. Then just keep heading north. Avoid populated areas and don't cross by the great lakes; select a crossing point in North Dakota or Montana.

Please keep in mind that while you are con-sidering becoming a hobo, it is done only to disappear. Being a real

When it comes to crossing boarders, plan it well so you do it in a safe place at a safe time. An isolated forest spot between 4:00 and 5:00 a.m. should do it. Constantly check for drones; if you see one don't run, just freeze.

hobo and homeless is a very hard life fraught with dangers, both in an urban and wilderness environment. What I am suggesting is a mix of hobo, hill walker, and survivalist. Time and planning should always allow you to return to society decently dressed and with some finan-cial support to see you through until you are working again.

As I originally stated, not all of us will have the financial means to disappear and never be found again; if this is you, then this section will give you some help. It is an exit from life strategy and if you stick to the rules, it will work. In essence, you simply fall off the planet, but in doing so you must be prepared to live a caveman existence, at least for some years. With that said, there is one great advantage in adopting this method: you will be self-reliant.

While my first thoughts are to send you out into the endless Canadian tundra or the jungles of Borneo, there are a few more places that offer sun, sea, and sand. Saint Helena is a small, 122–square–kilometer volcanic island in the South Atlantic Ocean. It has been used as a prison by the British and French. The British shipped some 5,000 Boers to the island from South Africa during the Boer War.

Or you could try Easter Island. However, both are populated, and your presence would soon come to be known, so your cover story would need to be very good.

The great thing about walking into the wilds is the lack of people, as well as ATMs to tempt you, and very little else to give your existence away. You must also consider that being a hobo and making your way on foot for the majority of time, while challenging, can also be an adventure. Additionally, you only need to do it for a certain length of time before you can rejoin civilization (unless you're a wanted fugitive). If you're a normal person who just wants out, living as a hobo for several years will drop you off the radar. When you return to civilization after, say three years, you could technically start again. Simply drift into a small town and look for work while pretending you have had a mental breakdown and cannot remember who you are or where you came from. Keep talking about living in the woods—at some point society will have you back and give you a name and a set of identity papers. Trust me, it'll work.

Although it's a bit isolated, you could always take a long boat deep into the jungle.

Choosing Where to Go

Should you opt to disappear into the wild, then you must first make sure you know where you are going. Second, be prepared for when you get there. Ideally, you should choose a wide open, uninhabited place where wild animals still live in their natural habitat and, most importantly, where a person can survive. Believe it or not, there are still some places left on the planet where this is possible. As man is a tropical animal, my list of preferences starts with the hotter countries, but places such as Canada should not be ruled out. Despite the cold, they are abundant in wild animals, fish, and edible plants. However, not everyone wants to live like an Inuit. I have written briefly about some of the best places I have been and remained for a long period of time; these are just for reference and you should choose a place best suited to your own plan.

Asia

Borneo is a really safe place. I have lived in the jungles of Borneo for many weeks, and what a pleasure it was. There is abundant game and fish and fruit on the trees just waiting to be picked, as well as ample material to construct a warm, protective shelter. You will come across people, but the communities are scattered and the locals are very friendly. If you can find your way to Kuching in Eastern Malaysia, you should be able to get a bus or a lift to the Batang Ai Dam and get a boat over to the riverhead. This is not as difficult as one might think, as the Hilton chain runs the Batang Ai Resort (one of the best and most luxurious resorts on the planet) and they run a regular boat from the road head across the dam to the resort. From here, you can get one of the locals to take you up river on a long boat.

The further you go up river, the less likely you are to see habitation and eventually you will come close to its headwaters and the Indonesian border. Scattered around this area are a number of small long houses, which are home to the Iban tribes. While in the past these people were known as head–hunters, I can assure you that today they are a highly educated people with warm hearts. The Iban are natural hunters, farmers, gatherers, and a perfect example of living in the isolated jungle.

Living with the Iban is not as bad as it sounds. They are a happy, well educated people, despite their forefathers being head-hunters.

Northern Thailand is also a great place to hide away. While much of this is farming country, the locals take little notice of foreigners. You would be well advised to get a job on a farm working for your bed and board only (most don't even pay their own family for working on the farm). The food is simple, the bed may well be shared, and the work hours in the fields long, but weather is good, the people are extremely friendly, and you will have the advantage of other humans to speak with.

> **Author's Note:** I spent about three months in Northern Thailand trying miserably to learn the language and failing (languages have never been my strong point). I arrived by bus after a twelve-hour journey—which still gives me nightmares—and arrived in a small village several miles north of Lam Plai Mat. I stayed for two nights in the local hotel before asking around for work. I offered my services for free in order to study the language. I had several immediate offers, but one farmer spoke English, and I decided to work on his farm. I literally dropped my rucksack at his home (shack) and went directly to the field. The job varied, but mostly involved digging ditches. It was hard, but I was young. At night I would shower outside—in full view of the whole family—then we would all eat at the same table. Meat was scarce, but there was always plenty of rice and locally grown vegetables. Being the only white person in the village made me a bit of a celebrity; the village was full of beautiful young girls, all of whom wanted to marry me. I have traveled much of our planet, but I can honestly say there are no kinder, hardworking people than the farmers of Northeastern Thailand.

South America

If I am honest, while the jungles of Belize offers similar life support as those of Brunei, the animals are a lot more dangerous, as is the vegetation. There are lots of small pigs running around that have tusks and are not afraid to attack humans. Likewise, some of the trees have acid for sap and cutting

There are always the odd places where it would be possible to hide out—the Mennonites in Belize are a perfect example.

them can make you blind. Water can be hard to find, but it does rain at 3:00 p.m. every day.

The one good thing about living in Belize is that everyone minds their own business. Belize is made up of numerous peoples: Creoles descended from Scottish buccaneers and African slaves make up a quarter of the population. Another third of the people are Mestizos—a combination of Spanish and native Maya—who tend to populate the inner jungles. However, in 1959, a religious group from Russia arrived and set up a community very much like the American Amish. They call themselves Mennonites after the Dutch Reformation Anabaptist Menno Simons. They live in strict accordance with the Bible and work very hard at farming the land; they have built a reputation among Belizeans for honesty and thrift.

If you believe in God with enough conviction and are not afraid of work, it might be possible that they will take you in. You could then stay a few years until the dust back home has settled and, if preferred, go elsewhere. While alcohol and most modern technology are forbidden (mobile phones, iPads, etc.), they do eat really well. If that doesn't work, you might try and get a job logging—cutting trees down in one of the many jungle clearance operations going on down there—but it's extremely hard and dirty work.

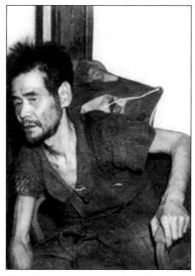

If you think it's not possible to live in the jungle for any length of time, you are very much mistaken. Shoichi Yokoi, a lance corporal in the Japanese Army during World War II, remained hidden in the jungles of Guam for some thirty years. He remained in his hide, believing that Japan was still at war and

Shoichi Yokoi survived in the jungle alone for 30 years.

that his former comrades would one day return for him. Then on January 24, 1972, Lance Corporal Yokoi, now fifty-seven years old, was discovered by local hunters.

He had survived by eating toads, river eels, and rats. He made traps from wild reeds to catch his food and discovered all of the edible plant life that surrounded him. Due to his military bearing and the desire to remain hidden, he had dug himself an underground shelter that was camouflaged on the surface. This was his home. Yokoi was not impressed with modern day Japan and his most prized possessions remained his eel traps and the simple tools he had made.

Upon his repatriation to Japan, he became a celebrity and in 1991 fulfilled his desire to meet the Emperor. Lance Corporal Shoichi Yokoi died in 1997.

The Canadian forest would be one of my first choices of places to disappear. Having spent many happy years surviving in God's creation, the only problem is the winter.

North America

Canada is such a beautiful country rich in just about everything a human could ever want when it comes to survival. The only problem is the weather. It gets very cold in the winter, and you will need resources to carry you through those long months, although game can still be found and ice fishing is possible. If you choose Canada as the place to spend your hobo years, then I would suggest you do a little research. You can overcome the problems of the winter if you can locate a small holiday cabin that's not used during the cold months.

As large as Canada is, you will at some stage come into contact with civilization. When you do, it will be to your advantage to play the role of an intrepid explorer rather than a hobo. This is especially true when entering a small town for the first time or passing through a middle–class residential area with neighborhood watch posters everywhere. Look clean and tidy, smile, and have a good excuse if challenged as to why you're on the road.

The American/Canadian border is clearly marked, but you would be better off crossing in an isolated spot.

Canada is a big place and you have a lot of choice of where to go. Personally, I would head north of the Manitoba lakes or to the foothills of Alberta, around Hinton. The latter is a paradise of fish–filled rivers, moose, and grouse just sitting there waiting to be picked up.

Author's Note: Always skin your grouse—don't just pluck it. Clean it out well and wash with clean water; use the guts for fishing bait if there is a river close.

Dotted around are many summer cabins, some big, some small, but all will offer you protection against the cold of winter. That said, you'd be an idiot if you did not prepare yourself properly for the Canadian seasonal conditions.

You might find a few rangers or people from the Department of Natural Resources lurking around, but by and large, providing you don't start a forest fire or shoot a gun within their vicinity, you should remain hidden. The good thing is, if you get injured, you can always signal them for help.

If you intend to live in an urban environment, you can very much forget about learning any survival skills: hand-to-hand combat training is better suited. You will get aggravation, not only by the police, but also by locals who don't want you around and even your fellow down-and-outs who simply want to rob you. Being a hobo in the city automatically sets society against you. Even if you manage to get into a homeless shelter, you risk lice and having your meager possessions stolen while you sleep. Worst of all, in some countries you might wake up to discover your kidney is missing. Park benches, alleys, store doorways, and any place in the city are bad news, and you simply cannot hide from the abuse. Forget being an urban hobo, learn some survival skills.

Places I would avoid are anywhere where there is civil conflict or a war raging, which basically rules out most of Africa and the Middle East. That's not to say all the countries of these continents are bad, but even the safe ones are far from ideal if you want to disappear. The desert is not a good place to survive, and the jungles of Africa harbor a lot of dangers. Both America and Europe have a lot of rules and regulations that make trying to hide difficult, although not impossible. However, Spain and France have large land areas in which to disappear.

Be wary of abandoned homes close to a small town or city, as it is a good bet they may be used by other homeless folk or, even worse, the drug fraternity. If you should find one that is fairly well hidden and you intend to occupy it for a few nights, remember to cover the windows at night to prevent showing any light from a fire or lamp you might have, i.e. do not draw attention to yourself. Also make sure the room is well ventilated if you start a fire, and check that the chimney smoke is not showing too much.

Hobo Help

Don't kid yourself—you are not a real hobo. You are merely using this as a disguise in order to disappear. While posing as a hobo is a good cover, there are many risks involved. First, if you stick around populated places, you will come to the attention of the authorities—something you should avoid unless looking for part-time work. Secondly, being alone in the wild has a lot of drawbacks. One major disadvantage of this form of disappearance is your health: If you get sick or are in an accident, then you are on your own. Breaking a leg in an isolated part of the Canadian tundra is definitely not the best way to end your days. You could, of course, make some preparations for this or use a satellite tracking device and call in a rescue. If you intend to go deep into uninhabited country alone, then I would suggest as a matter of course that you take a satellite tracking device with you. There are several on the market, some that will instantly report your distress signal together with your location. Others, which are similar, can be used in conjunction with your mobile phone to call several of your personal family or friends. The two main problems with this is that you will reveal your presence and have to explain your disappearance, and you will need a power supply to keep the device active.

Barry Davies

If you do go deep into the wilderness, always have a backup device for safety. Take a satellite location device with you and enough power to keep it working.

Travel

Traveling as a hobo is one of the first problems that you're going to have to face. If your disappearance destination is a long way off, you will need to carefully plan your route. Gather as much intelligence as possible about how to get to your final destination. Try to gauge how far you will have to travel, then break this down into various legs and plan how you will cover each section of the route.

You may have enough money to go part of the way by train or bus, but always plan for using your feet and walking—let's face it, you are in no hurry. Essential items, such as a satnav, compass, and detailed map, along with a good knife, flint, and steel should be securely attached to your person, preferably with a line so that you never lose them. Keep the use of anything that requires batteries to a bare minimum, and use your satnav only when you are lost or wish to confirm your position.

Before you start, it is important to work out a travel routine. This can always be changed once you get to know the country you're traveling through. The pace should be steady and unrushed, with a break of five to ten minutes at least every hour. Use this break productively, not just for resting, but also to evaluate your progress so far and to consider the next part of your route. If you're on a lonely dirt track, you may be able to get a ride with a passing vehicle. While this does present a small security risk, if your cover story is good then the risk is worth it.

When walking, always start the day as early as possible. At the end of the day, aim to pitch camp well before sunset so there will be enough time to eat, repair, organize, and settle down carly. Make sure you do not go beyond your physical limits. After walking all day, a good night's sleep will refresh you.

Maintaining personal hygiene while traveling is very important, especially if you want to catch a ride with passing traffic. Likewise, you may want to stop off to eat at small villages or towns. No one likes a smelly

Staying clean will help reduce the chance of disease. A lightweight solar shower in your rucksack is ideal.

person, and it's easy to keep clean and tidy. If you consider that your feet will be doing most of the work while you are traveling, it would be wise to take care of them—prevention is much better than cure. At the end of every day's march, remove all footwear and wash your feet. Also, wash and clean your socks and hang them on the back of your rucksack to dry.

Hitchhiking

It is possible to hitch a ride in almost any country in the world, but there are a few basics that you'll need to understand. First, someone else is offering you a lift in their vehicle; they will have expectations, doubts, and a little bit of anticipation as to if they should offer you a lift or not. You can help them decide by your appearance, your manner, and a clear indication of what you expect.

- Dress casually to look more like a camper than a tramp.
- Make sure you are clean—long hair and a scruffy beard will not entice anyone to stop for you.
- Smile when you put your thumb and don't get aggressive if the first half-dozen vehicles pass you by because the next vehicle might give you a lift.
- Make your intentions clear by having a readable sign with your destination clearly visible.
- If the vehicle does stop for you, be courteous and briefly explain where you're going.

In some cases, you may have to sit in the back of an open truck; make sure you have some weatherproof clothing in your rucksack.

If you are hitchhiking over a long period of time, you're almost guaranteed to be stopped at some point by a police patrol car. Most officers will simply be curious about who you are and where you're going. Be polite and never argue or be disrespectful to the officers, as this is a sure way to find yourself getting a lift in the back of the squad car. It is also a good ploy to ask the officers how far it is to the next town or if they know of any registered campsites close by. This will help delay any suspicion that you are a total down and out and might be a threat to their local community. In some countries, you will need to have some cash on you and be able to show you are not a vagrant.

Hitchhiking is a great way of moving around, but remember, people will want to talk to you and they will remember your face and dress, etc.

While you're hitchhiking, it's always prudent to look out for a good, quiet campsite that is free and where you will not get hassled. I would suggest that from about 3:00 p.m. you start looking for anywhere that offers shelter, a degree of warmth, and some privacy. If you're on the road, you'll find that many bridges or dry culverts offer excellent shelter. However, if these are too close to the highway, it can be extremely noisy, which will result in restless sleep. You would be better to walk some distance away from the highway, possibly into a forest area, and find a clearing close to freshwater where you can safely make a small fire.

If you plan to hitchhike over a long period of time, I would suggest that you plan your financial resources so you spend at least two nights a week in relevant comfort. This means sleeping on a proper bed, showering, and basically being warm and comfortable. There are many cheap motels, hotels, boarding houses, and trailer parks that offer overnight accommodation. If you have your own tent, I would suggest you utilize any campsites that offer full ablutions where you can clean yourself up.

If you have no money and are feeling a little jaded, you may try and seek assistance from any one of the many organizations that help the homeless. In America, one of the largest groups is called Volunteers of America (VOA). This group does invaluable work helping both individuals and families who've become homeless and are in need of assistance. A directory of all the local offices where you can seek help can be found at voa.org/Local_Office_Directory.

Jungle Travel

My second choice would be to disappear into a jungle region, as the jungle supplies just about everything you will ever need. You can always pass yourself off as a survivalist testing out various survival techniques; the locals will love you and will want to show you every edible plant and how to catch game.

Barry Davies

If you intend going into the wilderness of the jungle, it will help your case if you learn something about wilderness food.

However, if you have never been in the jungle before, it can appear like a bit of an obstacle course, which presents its own particular difficulties. Vegetation tends to be very dense and having to cut a path through will be an exhausting and slow task. In fact, when traveling through this type of terrain, normal progress would be about 1 kilometer per hour (approximately 2/3 of a mile), or about 5 kilometers a day (approximately 3 miles). Therefore, whenever possible, use alternative, easier routes even though it generally means you won't be traveling in a straight line. Native paths, game trails, dry watercourses, rivers and streams, or ridge crests will all provide you with a slightly easier passage. However, bear in mind that animals will also use these paths and trails—especially at night—so avoid them after sunset. It is safest to stay in your camp during the evening hours.

Be aware of the additional threats to your health and survival by biting insects and leeches. Take the necessary precautions and you should be able to limit the discomfort and damage they cause. In wet areas and after rainfall, check exposed areas of skin regularly for leeches. If you should find one, simply pull it off—they do not normally leave their heads in your skin to go septic as some people think. In jungle regions, always check bedding, packs, clothes, and especially boots before putting them on as there can be an unpleasant surprise hiding inside. By and large, I have always found the jungle very comfortable, and given the correct protection and a mosquito net, you will have a good night's sleep. Remember: never sleep on the jungle floor, as you will get bitten to pieces.

Choice of Route

When walking through a forest or jungle, remember that the shortest route is not necessarily the easiest. Always keep in mind that you are disappearing, so avoid major areas of population, main roads, etc., and use small tracks or pathways. In areas that are likely to be populated, travel—if you can—by night, and use the day to observe and rest. However, in less populated areas, it is easier and safer to travel by day and rest at night. If you can, choose to follow a trail along a high ridge rather than a route that takes you through a valley. Valley routes generally contain more obstacles, such as thick undergrowth and possible river or stream crossings, and most populated areas are in valleys. Another hazard is swamps or marshy ground, which are at best hard going and at worst dangerous to navigate. Ridges also tend to provide a visual advantage that will make it easier to keep your bearings. A ridge might be oriented in the direction you are traveling, and if so, this is very fortunate. More often than not, however, it is likely that the route of the ridge will head off on a totally different bearing than the one you wish to follow. Even so, it might still be worth following the ridge for a short while, keeping an eye out for a suitable alternative ridge, valley, or river crossing.

Contours offer a useful halfway measure between ridge and valley floor. A trail that follows such a contour might take a longer route than a ridge top, but it will mean that less climbing has to be done. Basically, your decision will have to be made based on careful observation of the terrain.

Mountains—especially the lower slopes—offer the walker many possibilities for shelter, food, and water. Their drawbacks are that traveling might be more difficult and the survivor may be at greater risk of injury. There are four basic rules to remember when traveling through mountains:

Be careful when moving over desolate terrain and use a flotation aid when crossing water.

Barry Davies

- Avoid any loose rocks when climbing and always make sure you have three points of contact.
- Always make sure that it is possible to climb back down again if you need to.
- Stay out of snowfields, glaciers, and overhanging snow.
- Never try to cross an obstacle that could cause your death, i.e. crossing a swollen river or climbing a sleep cliff face.

Obstacles

Natural obstacles are things such as rivers, swamps, and mountains; you are likely to encounter many while you are traveling through the forest or jungle. Some will stop or slow you down, while others will dictate a change in your direction of travel. Streams and rivers, for example, always run downhill, and they tend to lead to areas of human habitation.

That said, rivers and their immediate surroundings contain many resources useful to the walker, such as food and drinking water. There are a few hazards—such as mosquitoes and other biting insects—that will be located close to water, and heavy rain can cause the river to swell quickly. Attempting to cross a large or flooded river is a very dangerous pastime; never cross any water unless you are 100 percent sure it is safe to do so.

Don't attempt to climb high-sided mountains or rock faces; look for an easier way around. Be observant for areas of rock side; again avoid such areas where at all possible. It will be beneficial in the long run if take your time at each obstacle you reach and look for an alternative route. Likewise, if you start to get stuck in a swamp, turn back and regain firm ground. You have all the time in the world, never risk getting injured.

Hobo Dress

Remember, you are not a real hobo—you're just pretending to be, and the best way to approach this is as a long-term camper who is hiking his or her way around the country. Therefore, you should prepare and equip yourself accordingly. This means dressing correctly and having the right clothes for the weather conditions: good footwear as you will be doing a lot of walking, a good sleeping and camping system, and enough tools and devices to make life bearable. You should also carry enough food and water to see you through at least a week.

This image is of a hobo in years gone by. Dress as if you are a hiker and not a tramp, this way you will avoid a lot of aggravation from locals and the law.

Clothing

If you intend to be on the road for long time, then you will need a certain amount of protective clothing. Remember most of what you have will be carried in your rucksack. Having been in the military for many years, it is easy for me to pack a rucksack that will last me for at least three months. However, because of weight and bulk, you will need to minimize everything that you have to carry.

The human body functions best at an internal temperature of 96° F and 102° F. Above or below this and you can start getting into trouble. It's easy to get cold and it is equally as easy to overheat and sweat; the ideal is to operate without doing either. The best way to do this is using a layer system.

The layer system is simple: the colder you are, the more layers you need, some of which can come off as you get warmer. You will need a layer next to your skin—this should be loose fitting and made of cotton. The next layer should seal at the neck and wrists if required. In very cold conditions, you may put on a third layer such as fleece, but this can cause the body to overheat. Your outer layer should be wind- and waterproof.

My advice would be to wear some form of protection for your head and hands. The heat loss through your head is particularly high, so a comfortable or woolen cap is required. You cannot work properly if your hands are cold, so always have a good pair of gloves; a set of cotton inner and waterproof outer gloves are perfect.

The amount of outdoor clothing available today is vast, so you should have no problem finding the right range that will suit your needs. Look after your clothing while alone in the wilderness; keep it clean and in good repair so that it lasts you and functions as it should.

Boots and Socks

If, due to your financial cir-cumstances, you are forced to become a hobo and you envis-age doing a lot of walking, then you're going to need very good shoes and socks. The right foot-wear is like an old friend; it

I cannot overstate the need to have good boots and look after your feet enough. As every soldier knows, if they fail, you fail.

never rubs you up the wrong way and will support you when the going gets tough. The fact that we carry our entire body weight on two feet instead of four—like most animals—means that we place a lot of pressure on our feet. This is especially true when we are also carrying the additional weight of a heavy rucksack or walking over rough terrain. It is, therefore, important that you look after both your feet and your boots.

As a general rule, you should choose a pair of walking boots that combines lightness with adequate support and protection. Boots can come in a bewildering array of styles and suitability for different terrains. Good boots are often quite expensive, so most hill walkers can only afford to possess one pair. Invariably this pair then has to be able to cope with all kinds of conditions.

Socks are also important. Lightweight fabric boots are designed to be worn with a single sock, while heavier boots might require two pairs. In the latter case, the inner sock should be lightweight wool or silk for warmth, while the outer sock should have thickness to cushion the foot. Here are some of the things one should look out for when purchasing a new pair of walking boots:

- Always wear the same type of socks you would go walking in. Make sure your toes are not touching the end of the boot. A good fitting boot should feel comfortable but not restrictive.
- The boot backstay should protect and support the boot, as should the heel corner and toecap.
- The boot should be high enough to protect the ankles, with a padded scree collar and a bellows tongue to protect against water and debris.
- The insole and lining should cushion and support the entire foot.
- The boot should be waterproof.
- A good grip is essential, especially on wet, slippery rocks. Try to avoid PVC; choosing a rubber star–patterned sole will give you much better grip. The sole thickness will depend

upon the sort of terrain the walker proposes to be exploring. General hill walking does not require an extremely stiff sole, unlike in mountaineering. Try twisting the sole to see if it is flexible; if it twists easily it will not give much support during a fall.

- Try to stand on an incline, or tap the heel and toe of your boot. If the toes feel trapped get the next size up.
- Never purchase new footwear if you have any form of foot ailment, such as ingrown toenails or corns. Wait until they have been treated.

Don't buy a pair of new boots the day before you intend to disappear, as it takes time to get used to them. Start off by wearing them with the laces slightly slack and always make sure the tongue is neatly flat against your foot. Wear them around the house or go for short walks; this should iron out any hot spots before you disappear. Always clean mud off them at every chance, then wash and polish or spray them. Any detachable insoles and wet laces should be removed and dried thoroughly using either the sun or other heat source. Beware though of putting wet boots too close to an open fire, as leather tends to crack when it dries too quickly. Instead, dry them out by stuffing them with an absorbent material, such as paper or tissue, and leave them in a warm place. Once the boots are dry, apply several layers of a good waterproof compound, making sure that each layer is well rubbed in. Regular and careful care will prolong the active life of your boot.

Author's Tip: If you intend to walk through deep and heavy forest such as you find in the Canadian tundra, then I suggest you also use gaiters that cover the entire boot and upper leg, as they will help keep your boots dry. Gaiters should also be worn in snowy conditions, as they will keep most of the snow from going down your boot.

Rucksack

In addition to your clothes, boots, and socks, the next important item when doing any long-term hiking is a well-constructed rucksack. There are many to choose from, and the earlier in your pre-disappearance phase you purchase one, the better off you will be. Like a pair of boots, a new rucksack needs breaking in. Your rucksack will be come your mobile home, so it should hold everything you need to live outdoors while remaining dry and comfortable. Most Special Forces soldiers will tell you about how to pack a rucksack, and most will do it right. You should make sure that the contents of the pack are relevant to your needs; that is to say, don't carry weight you don't need. Make sure items you need during the day, such as for cooking and drinking, are accessible without having to empty the rucksack. Likewise, items you use just once a day, such as your sleeping bag and shelter, should be in the bottom of your rucksack.

Shelter and Sleeping Equipment

You're not always going to get a bed to sleep in, and it is possible, as a hobo, that you might need to doss down on the side of the track for the night more often than you think. This being the case, I would recommend a good modern shelter and a sleeping bag.

The best shelter, which can be erected in seconds, is a one–man, military–type Special Forces shelter. They are fully protected against the elements and keep a low profile. There is nothing like it to provide you with a good night's sleep no matter what the weather throws at you.

Barry Davies

Life against the elements can be easily solved by carrying a small one–man tent similar to those used by Special Forces.

Add to this a good quality sleeping bag and a foam insulation mat, and you are going to be snug as a bug in a rug. The good thing about them is the size and weight. Both the shelter and sleeping bag will roll up very small, and the insulation mat is extremely light. If I had to endorse any bits of outdoor camping equipment in order for you to disappear as a hobo, these three would be at the top of my list. As any soldier will tell you, a good night's sleep on a cold and wet night is worth its weight in gold.

As a hobo with no proper sleeping system, you will need to look for every opportunity where you can rest your head; avoid the wind and rain and keep warm. Both urban and rural terrain offers a multitude of places to sleep in comfort, but your priority should be on finding a place that offers your some seclusion and security, as well as protection from the elements.

Survival Kit

Going it alone—whether it's walking across the back roads of Canada or in the jungles of Borneo—presents many problems. In the life you have just left, these problems are normally fixed by others, but when you're alone, you will have to fix them yourself. It might be a simple thing, such as a broken zip on your jacket, but if it's not fixed the cold will get in. You may fall sick from eating something bad; you will need to fix this yourself. When you're alone, you need to cater for most emergencies, even the most basic ones. In order to do this efficiently, you will need a survival kit.

Every item within the survival kit will need to be of use, but its size and weight also need to be taken into consideration. This will mean that sometimes you will need to make difficult choices on what to include and what to leave out. Ultimately, each item must increase your chances of survival, bearing in mind that you may have no other initial resources. The components should be viewed as a primary tool kit that will help you to capitalize on your own survival skills. A selection of items that could potentially be included in a survival kit is listed below. The notes will help you to decide on their usefulness in different situations. However, the final choice will be dependent upon your personal preferences, skills, and the location where you intend to disappear.

The Special Forces Survival Kit does a bit of everything.

Barry Davies

Remember, your survival kit is not just about hunting knives and fire–making tools. While these are essential, you will also be eating every day. Always pack a good supply of salt, pepper, and curry powder. Additionally, a large supply of multivitamins will help keep your body from deteriorating, especially if your diet is poor and your meals are infrequent.

Summary

While disappearing as a hobo can be done almost instantly, a little planning will always improve your chances of never being found. You will have the opportunity to either stay in an urban area or travel into the wilderness. While the cities and towns offer you more opportunities and cater for the homeless, the chances are you will be at risk. You will always be safer in the wilderness and, if properly equipped, you can survive.

Living the life of a hobo does not mean looking like a dirty tramp who everyone despises; keeping yourself clean and tidy will always enhance your wellbeing. It will also help keep you free of disease and prevent unwarranted illness. You will need to consider what clothing and shelter you take with you, as this can be the difference between being cold and wet or snug as a bug in a rug.

Likewise, the deeper you go into the wild away from civilization, the less man–made resources will be available to you; learn how to survive.

9

HOBO SKILLS

While I have no intention of turning this book into a survival guide, there are a few things that need to be explained. When you set out to challenge the great outdoors—in addition to selecting your equipment—you will also need a set of skills. These skills are designed to keep you comfortable, well fed, and most importantly, alive. Now, you could simply go out and buy one of my several survival books, or one written by my friend John "Lofty" Wiseman, the grandfather of survival (*SAS Urban Survival Handbook*). But while books show you the techniques in principle, my advice would be to get yourself on a good survival course for a weekend. Apart from being fun, you actually get to learn the survival skills first-hand. At the very worst, you should learn the basic skills of lighting and caring for a fire, how to construct a shelter against the weather, finding food that's safe and edible, and gauging the weather before you set off for the day. Knowing what to do should you get caught out in really bad weather or if you encounter an emergency could help save your life.

No matter where you plan to go in the world, if you intend to live the life of a long–term camper or hobo, you will need some basic skills such as understanding the weather and where to find free food and clean drinking water. Starting and maintaining a fire to its maximum without setting fire to the forest and looking after your health are also very important. Then there are the skills that just might help if you should need to temporarily occupy a deserted home, or get yourself a lift by borrowing a car. These could include breaking and entering, or how to hot–wire a car, etc.

Weather

In many places, the weather can be unpredictable, and this is doubly so on hilltops, mountains, and in forests. Weather can cause many problems and potential dangers for someone alone, so you must be aware of sudden changes at all times and be prepared to act accordingly. Although we cannot change the weather, we can, to a certain extent, predict it and recognize its approach.

On the open road or in the wilderness it is well worth learning a little about clouds, temperature, and the wind. This way, you will have an idea of what weather to protect yourself against.

I have found that weather predictions can be very inaccurate on occasions, and to compensate for this I have found a method of anticipating any immediate danger from the weather by observing the sky and clouds. A clear sky with high clouds will indicate a clear and sunny day. A dark sky with low clouds will normally indicate rain. It is simply a matter of gauging the degree between the two. I do this by looking toward my direction of travel and trying to estimate the height of the clouds, color of the sky, and wind direction. With a little practice, one is able to anticipate the weather conditions for several hours ahead. Look at the sky and see which way the clouds are moving; they can tell you a lot about the coming weather.

- Clouds moving in different directions almost certainly herald bad weather.
- Altocumulus clouds, those which look like mackerel scales, also mean bad weather is on the way.
- Cumulonimbus clouds early or developing throughout the day can mean chances of severe weather.
- Cirrus clouds, high in the sky like long streamers, mean bad weather within the next thirty-six hours.
- Cumulus towers indicate the possibility of showers later in the day.
- Clear skies mean the weather is fine, but it could get cold at night.

Actually, looking up at the sky and trying to determine the weather is a good skill to have anytime. It's easy to do, costs nothing, and is something to occupy your mind. Today you can get a wide variety of mobile phone apps that will provide you with a fairly accurate prediction no matter what part of the world you are in.

Author's Note: Regular soldiers who want to join the British Special Forces (SAS) must go on what is known as "selection." Selection takes place twice a year; once in summer and once in winter. Part of the selection is a series of marches with a heavy rucksack over mountainous terrain. The marches are commonly referred to as "test week," at the end of which candidates must complete a 40-mile march with a 30-kilo rucksack (66 pounds), plus a rifle.

One candidate, Captain Carnegie, did not make it. He was found in freezing weather on the 2,907–foot Corn Du peak on Saturday, January 26, 2013. He had collapsed and died from hypothermia. This is not the first time soldiers on SAS selection have succumbed to the weather in the Brecon Beacons. This incident alone demonstrates how the weather can beat even the fittest of men.

Survival Food

There are two sources of food from the wild: plants and animals. Animals will usually supply richer food than plants, but it costs time, effort, and energy to hunt or trap them. The sources of wild animal food include mammals, birds, reptiles, fish, crustaceans, and insects. Any of these can provide food that, pound for pound, has much higher food value than most material derived from plants. They do not provide it willingly, however, and have to be hunted, trapped, or caught. To do any of these, information and skills are required. If you're near habitation, farmers' fields are a great supply of vegetables, but always check that they have not been sprayed with some poisonous insecticide.

Food from plants, on the other hand, is usually readily available, except in extreme conditions or locations. Plant foods (including roots, leaves, berries, fungus) might not provide a fully balanced diet, plus they can be relatively low in food values. You may have to eat greater quantities than normal to meet your body's requirements. However, plants are sustaining and they are easily obtained if you know where to look and what to look for.

Obtaining food from wild plants is a skill that has to be learned and practiced. Well under one half of all plants are edible, and most of them only in parts. Some are poisonous and will make you ill or even kill you. Therefore, knowledge and skill are needed to take advantage of nature's bounty. If you know what plants to look for in your location, you should normally be able to find enough food to keep yourself alive.

Meat

There is enough food on our planet constantly being renewed to feed us, so all you have to do is to recognize what is easy to catch and what is digestible. Take ants for example. Ants are the Earth's most abundant insect species, and their total biomass (weight) is greater than that of all the mammals on the planet combined. While ants might not look very digestible, you should take advantage of the "protein soup"

Worms and snails cannot run—all you have to do is pick them up. Both make good eating if cooked properly.

recipe on the next page. Basically, if it walks, crawls, swims, or flies, then it's edible (well, almost all of it). Below is a short list of some of the millions of insect and mammal species that live on our planet. These include:

- Ants
- Worms
- Grubs
- Termites
- Cockroaches
- Bees and Wasps
- Snails and Slugs
- Hedgehogs
- Rabbits
- Rats
- Snakes
- Wild Cats

- Wolves and Wild Dogs
- Pigs
- Deer
- Bears
- Goats
- Seals
- Penguins
- Birds
- Camels
- Crocodiles and Alligators

- Lizards
- Kangaroos
- Cattle
- Horses
- Monkeys and Apes
- Turtles and Tortoises
- Frogs and Toads
- Crab
- Shrimp
- Shellfish and Fish

Protein Soup

If you should find yourself with insect or small animal food that is not palatable, you can always turn it into protein soup. Take ants for example: put as many as you can get into a tin and place the tin on the fire so that it dries the ants out, roasting them. Once roasted, use a spoon or stick to crush them onto a fine dust. You can then add a little water and all the legs, shell, and bad bits will float to the surface where you can scrape them off and discard. What you have left is a brown soup. Bring this to the boil for several minutes and then drink. Most insects are high in nutritional content, so you can make protein soup from grasshoppers, grubs, snails, worms, etc. You can try adding some recognizable plants—leaves and roots—to give your soup some body. Nettle leaves or dandelion roots are a perfect example. If you find that too many insect shells remain in your soup, try straining them out using a clean sock.

Hunting Weapons

While I do not advocate taking a weapon into the wilderness, I would suggest that if you envision some long-term survival, then you can take a powerful air rifle. It is not a particularly good idea to carry it with you all the time though, just in case you get spotted by a ranger. Instead, keep it in a safe and dry place for the days when you want to go hunting. The noise from an air rifle is not very loud so you should be okay. Do NOT try to shoot larger animals, as you will not kill them with the average air rifle and your pellets will simply cause them pain. While an air rifle offers accuracy, a crossbow offers killing power; enough to take down large game, such as a moose. The same rule applies: Only take it out when you go hunting.

Eggs

I mention eggs here because they are one of the most convenient and safe foods you can eat. The wilderness—especially the Canadian tundra—is home to millions of birds, both big and small. Find their nesting ground and you will find yourself a feast. Eggs have been with us since man first walked the earth, and they still come in the original packaging. They are simple to cook in a variety of ways, and if hard-boiled, provide a convenient and easily portable food reserve that will keep for several days. Covering eggs in fat or oil will help preserve the eggs for months (egg shells are permeable and it's the air that makes them go bad).

Author's Note: Most wild eggs are protected and I write here purely in the event of having to survive in the wild.

Eggs are the food friend of any wilderness traveler. They still come in the original packing, can be cooked and carried, and are full of nutrition and energy.

Best Foods to Carry

Apart from your water, sleeping bag, tent, and change of clothing, you will also want to take some food with you. As a hobo, you cannot expect to simply walk into a restaurant or store every day and buy yourself food. You will be out there on the open road or walking in the forest, so you are going to need a supply of food. One of the best ways to organize your food is to do what soldiers do and build yourself a ration pack, or better yet, buy some from a surplus store. A military ration pack contains all the things you need on a daily basis: food, matches, toilet paper, gum to clean your teeth, etc. In addition to ration packs, you should also consider a few other food products. This is my short list, and everything on it is designed to supplement any food you can get your hands on:

- **Oats:** High in fiber and complex carbohydrates, oats have also been shown to lower cholesterol. And they sure are cheap—a dollar will buy you more than a week's worth of hearty breakfasts.
- **Dried food:** It's lightweight and only requires water to rehydrate it. You can get both meat and fruit in a dried format.
- **Curry Powder:** As every SAS soldier will tell you, there is nothing better than a spoon full of curry powder to flavor your food; it makes just about anything edible.
- **Eggs:** You can get about a half dozen eggs for a dollar, making them one of the cheapest and most versatile sources of protein. If you have too many eggs, boil them, as they can then be used for a quick snack when walking.
- **Apples:** Apples are inexpensive, easy to find, come in portion-controlled packaging, and taste good. They are a good source of pectin—a fiber that might help reduce cholesterol—and they have the antioxidant vitamin C, which keeps your blood vessels healthy.
- **Onions:** They last forever in your rucksack and are great at bulking out a meal when times are hard.
- **Sardines:** They are an acquired a taste, but a can of sardines is relatively cheap. And the little fish come with big benefits: calcium, iron, magnesium, zinc, and B vitamins. And, because they're low on the food chain, they don't accumulate mercury.

Fire

The discovery of how to make fire was one of humanity's greatest advances. With the provision of shelter, it allowed humans to modify their environment, enabling them to survive in otherwise unsuitable climatic conditions. It is because fire has been such a vital part of man's history that it also plays an important psychological role in survival efforts. It is a source of comfort; the lighting of a fire is proof that a survivor can control at least some of the dangers that face him or her. It also provides a sense of achievement in that the survivor has replaced, in his or her emergency situation, one of the major elements that contribute to normal life. Even more importantly, fire is of practical use in many ways.

Fire will provide heat and light, together with the ability to cook food. With fire, water can be purified and medical equipment sterilized. Clothing can be dried. Signals can be made with the smoke when seeking help.

It is important to know how fire works in order to make fire the first time, every time. Fire requires three elements: heat, fuel, and oxygen. If any one of these elements is missing, a fire will not burn. When considering the supply of fuel, it is helpful to recall that fire is a form of chain reaction. Part of the heat generated by the combustion of any fuel is required to ignite the succeeding supply. The initial supply of heat available to start the fire is usually small—a match flame for instance—and lasts only a few seconds. It follows that the starting fuel, which must be set alight by such a brief flame, must be a material that ignites very easily. It must be some form of tinder.

Tinder must be dry, and it will ignite more readily if it is reduced to fibers, threads, or shreds. Any material that is suitable for use as tinder will burn quickly, and it is therefore essential that before attempting to set light to the tinder, you make certain there is a supply of kindling wood ready at hand.

Kindling should consist of small dry twigs followed by dry sticks, which will enable a small, hot fire to be built. You may then gradually add larger sticks until you have a fire that will burn long enough to ignite small logs. When such a fire has been established, even green logs can be added, since the heat available will boil out the sap before the logs burn. At first, however, the wood you gather should be dead and as dry as possible. Here are a few tips prior to lighting any fire:

- Collect and grade the fuel into tinder, kindling, and heavy logs.
- Do not pile kindling on to the fire too soon as this limits the supply of oxygen.
- Ensure that the fire is well-ventilated, so it will burn efficiently.
- Smaller drier wood will produce less smoke.

The heat required to start a fire can be generated in a number of ways. The easiest to use is an open flame, as from a match or lighter. Sparks from flint and steel or from an electrical source can also be used to ignite tinder. A magnifying glass or parabolic reflector can do the same in sunny conditions. Today, matches have mainly been replaced by butane lighters, but my advice is to carry a flint and steel on your person, as this will light more than 2,000 fires, wet or dry.

There might be an emergency where you will need to make a fire quickly in order to stay warm or to dry yourself. New outdoor fuels such as Fire Dragon cooking fuel can be used in such circumstances. Simply rip off a small piece and use your flint and steel to ignite it with the sparks.

Author's Tip: Always build your fire in a safe place and protect it from accidentally spreading or going out of control. Should you find yourself in a place for several days, one of the best ways to utilize fire is to build a rough stone Yukon stove. I have done this many times in my life, and still hold that it's one of the best skills I ever learned.

Building a Yukon stove is always a good idea if you are in the same place for any length of time, or the weather is poor.

Yukon Stove

By far the best use of fire for cooking and general purpose can be obtained from the building of a simple Yukon stove. If you are in one location for more than twenty-four hours, you should certainly consider the possibility of building this type of stove. It is also a secure way of providing heat and cooking.

Rocks, stones, and mud are used in its construction, with the tortoise shell as the basic pattern. At one side, you need to leave a hole for the intake of fuel and air, and there should be another at the top to act as a chimney. Two further refinements are very desirable: the first is the building in of a metal box or large can into the back wall as this will provide an excellent oven. You must remember, however, that food placed in the oven will be burned unless it is separated from the metal by small sticks or stones. If twigs are used, they will turn into charcoal after a day or two. You should keep them for use in deodorizing boiled water if necessary and other medicinal purposes. The second improvement possible is to use a large flat rock as part of the top of the stove. It can be used as a griddle for making oatcakes, drying leaves for tea, parching grain, and even frying birds' eggs.

One of the Yukon's major advantages is that it can be left unattended while you are working on other activities and you can return to a warm fire and hot meal. By covering the fuel and air intake with another stone, the rate of burning can be partially controlled. In wet weather, the oven enables fuel to be dried. Clothing can be laid over the outside of the stove and will dry without burning. You can also warm yourself without risk of being burned.

The Yukon stove normally takes about two hours for one person to construct. That is, if most of the materials are at hand. In my lifetime, I have always taken the time to construct a good Yukon stove, and it has made life bearable, in all aspects.

You must provide effective ventilation if you intend on using a stove or heater inside your shelter. This means two openings: one at the top of your shelter as a chimney and another close to ground level to admit fresh air. If you are in a heated shelter and begin to feel drowsy, you may be in danger of carbon monoxide poisoning. Get out into the fresh air, moving slowly and breathing easily and evenly. Most importantly, find and remove the cause of the fumes. If a group is sleeping in a heated, closed shelter, one of their numbers should stay awake on carbon monoxide guard duty.

It is better to rely on your clothing and other insulation

Barry Davies

Water is life. Without clean water on a daily basis your body will suffer. My suggestion is to use a military–type hydration system.

to keep you warm. Reserve the use of stoves or heaters inside the shelter solely for cooking. Only if you are positive of your safety should you extend their use to heating.

Water

If you are contemplating becoming a hobo and doing a lot of walking, then your first priority above all else should be water. Water is something we take for granted, but when you're walking out in the forest or jungle, it's not just a matter of turning on a tap or buying a bottle from the local store. Water is vital to your survival. In general, a human body, which itself is about 70 percent water, cannot survive without water for longer than three days in a hot climate and twelve in a colder one. In a temperate climate, carrying out a normal level of activity, the body requires a daily fluid intake of two–and–a–half liters. This requirement will rise in a hotter climate and with greater physical activity. If you wish to keep your body efficient, your minimum daily water requirements must be met. It is not only the quantity of water that is important, but also the quality. Contaminated or impure water will cause more harm than good and might put you at risk for serious disease.

My advice is always carry a water reserve with you and at each and every opportunity, make sure it's topped up. There are many ways to carry water, and the best, most efficient way is to use a rucksack that has a water bladder inside. This will allow you to carry up to three liters of water comfortably, and you will be able to drink without having to remove your rucksack. Ask any soldier: a combination rucksack and hydration system is perfect.

Filtering

Do not underestimate this risk. The disease-inducing and other harmful organisms contained in impure water constitute one of the greatest enemies to survival. If your only source of water is impure—or even suspect—do not drink any until it has been filtered and purified.

The first step towards making water fit to drink is filtration. This will remove creatures of any size as well as mud particles, leaves, or other foreign matter. Clean sand held in a short sleeve, sock, or cloth can be used effectively. A bamboo section plugged with grass also makes a good filter.

Always use purification tablets if available. Follow the instructions for use with care. If no purification tablets are available, boil the water for five minutes. Try to obtain a fast enough boil rate to agitate the water, as this ensures equal distribution of heat.

If the climate is hot and sunny, consider the possibility of setting up a survival still. It can be employed to purify water in the same way that it obtains it from the ground or vegetation.

Charcoal added to any purified water will help remove unpleasant tastes or smells if added an hour before drinking. Don't worry about any small pieces of charcoal left in the water when drinking it, as a small amount will do you more good than harm.

Of secondary importance to water is salt. The normal human requires about ten grams of salt each day to maintain a healthy balance. Sweat contains salt, as well as water, and this loss must be corrected. If it is not, you will suffer from heat stroke, heat exhaustion, and muscular cramps. The first signs of salt deficiency are a feeling of sudden weakness and a hot, dry sensation to the body. Resting and a small pinch of salt added to a mug of water will eliminate the feeling very quickly. In dry desert or sweaty jungle conditions, it is advisable to add a small amount of salt to your entire fluid intake.

Survival Shelters

As previously mentioned, you should really equip yourself with a proper shelter, similar to those used by Special Forces. However, given that you have undertaken to disappear into the forests or jungle (no one in their right mind would try to live in the desert or arctic), you will also need to know how to go about building a shelter.

How and what you build will depend on where you are and how long you intend to stay there. A shelter is there to protect you from the cold, wind, rain, and snow. It is essential to protect yourself against these, as each of them is a factor that hastens hypothermia. Exposure to any combination of them can rapidly produce deadly results long before any shortage of food or water would take effect.

There might be temporary shelter to be found among the natural features surrounding you. If you only require temporary shelter, seek it in or around trees, thick bushes, or natural hollows. If safe, make use of caves, rock overhangs, or any available natural shelter. Never waste time and energy constructing a temporary shelter or windbreak if nature or circumstances already provide it.

In the case of a more permanent shelter, the climate and terrain will always influence the sitting of a shelter and type of construction involved. There are, however, some general points worth keeping in mind when approaching the job. Choose a site that uses to its full advantage any natural cover from the wind. If no such cover is available, remember to angle the shelter so its entrance or open side is always away from the wind. Paradoxically, a hillside is usually warmer than a valley floor, even though it might be windier. Build the shelter as near as possible to a fresh water supply, to sources of building materials, and, very importantly, firewood. Any spot in a forest and near a fast-flowing stream can be the site of a very desirable residence.

In lowland areas, it is important to recognize the danger of flooding. On the coast, keep the tides in mind. In mountainous areas, make sure that the chosen site is not in the path of possible avalanches or rock falls. If in the forest, look around for fallen trees, as they may indicate an area of shallow soil. If the wind can blow one tree over, it could do the same to others nearby. For the same reason, isolated single trees are best avoided. On the other hand, the branches of an isolated tree that has already fallen could well provide a ready-made framework for a sound shelter.

Your Health and Injury

Once again, I do not intend to make this book on how to disappear into a medical manual, but it would be amiss not to mention and caution you about health hazards and dealing with an accident. This applies to all those who disappear; not just those living a hobo existence. You may have disappeared overseas and now find yourself living in a remote area where the medical facilities are nonexistent or primitive. If you are on your own and you become sick or get severely injured, it could be the end of all your plans. There are three basic things you need to do: look after you personal hygiene, carry a basic medical kit, and be able to recognize when you really do need professional medical help.

When accidents happen out in the wilderness and you cannot call 911, you will need a good medical kit.

Barry Davies

Hygiene

Bodily cleanliness is a major protection against disease, germs, and infection. A simple daily wash with warm water and soap is all that's needed. If this is not possible for several days, then at least make the effort to keep your hands clean. You should, if possible, wash or sponge your face, armpits, crotch, and feet at least once a day. Everything you eat—the power that keeps you going—is ingested through your mouth, so make sure you clean your teeth first thing in the morning and last thing at night. As mentioned previously, wash your clothing whenever possible and keep it dry. If you have no water, shake it in the air as much as you can. You are likely to get insect bites from time to time, but do not scratch them—this is the easiest way to get them infected.

Buy yourself a good medical kit and always carry it with you. Most homes have a medical kit on standby; you should do the same and keep one in your rucksack. Select your items with care and do a little research into wilderness medical emergencies. Learn how to assess the problem, and what your priorities are. Most importantly, learn what to do in the case of the following major problems:

- Bleeding
- Shock
- Fractures
- Concussion and Skull Injuries
- Burns
- Heat Exhaustion and Hypothermia
- Poisoning

Your disappearance plan might require you to be alone either as a hobo or when hiding overseas. In this case, it is even more important that you recognize when you need medical help. If you are alone and have a major medical problem but can still walk, then you should make your way as quickly as possible to civilization. If you are unable to walk, then you are down to relying on your back-up plan and will need to use either a mobile phone or, if you have no signal, a satellite emergency beacon. If neither of these are an option, you will have to fend for yourself as best you can. In many cases, if you can stop bleeding, immobilize a broken bone, and prevent shock and infection, you will have a chance. Keeping warm and dry, plus getting plenty of rest will all aid recovery.

Summary

What a person needs to survive comes down to decisions about what to wear to withstand the very worst of weather, the contents of their rucksack, and the supplies he or she will carry. Depending on the country or terrain being traveled over, it's up the individual to select any additional items, such as a mobile phone or satellite emergency beacon (remember, they all need power and you cannot rely 100 percent on getting a signal).

There is nothing better than good preparation, training, and reliable equipment to get you through most situations when you are alone. During my life, I have walked across the desert from the United Arab Emirates in the north to the city of Salala in Southern Oman—a distance of about 800 miles. I have also thrived living in the forests at the foot hills of the Rocky Mountains in Western Canada, again, totally alone.

Never be afraid of the wilderness; learn to live with it and not fight it. Learn to understand its good points and its dangers. Being alone can be a wonderful experience, but always make a mental note of what you will do if something goes wrong, as most dangers can be avoided or foreseen. Finally, if your life depends on it, forget your disappearance plan and make contact with the world once more. Anything is better than dying.

10

YOUR NEW LIFE

Once you have disappeared and finally arrived at your new destination, you will need to sort out a new life for yourself. You must live the lie you created, and play the part as if it were real. But no matter how happy you are and how good your new life is, you will need to keep an eye on the past. Watch for signs of anyone trying to find you and be ready if they do.

You should have already selected the type of life you wish to lead; this will very much depend on your own character and your financial reserves. Even though you may have visited your new destination country several times before, actually living there can be far different. For starters, you cannot simply live your life in a hotel, though you will more than likely initially book into one for a few weeks. You will need to find a home; a place that is safe and where you will not be discovered should people come looking for you. If your finances are limited you may also need to look for work. Again, if this is the case you should already have done your homework and know what is available to you, and what type of employment will help protect your identity.

Like most humans, we all need friendship, a partner, and a few home comforts. Once again, when choosing new friends, you will have to be careful, as friends ask questions about your background. If you are female, a new life can be difficult to establish, and you will need to be even more careful about selecting your friends. Nevertheless, the comfort of a local partner (I use the word partner here to represent a male or female sexual relationship) has many advantages: they will help provide cover, accommodation, and getting to know the dos and don'ts of the local society. Most of all, they will give you some warmth and comfort, which you are going to need as there will come a time when you will miss your family and friends back home.

Finding a New Partner

While some of us can survive alone for a length of time, most of us will eventually seek out a partner. The benefits of having a partner in a new country far outweigh not having one. The chance of compromise is very slim, especially if your cover story is watertight. Now the first problem you will have is your sex, age, looks, and sexual orientations. There are plenty of variations, so forgive me if I don't cover the whole spectrum, but instead write as if the person involved is a thirty-five-year-old heterosexual male. For everyone else, you will just have to work it out for yourselves.

A normal male of around thirty-five years old who is averagely fit should have no problems finding a good, honest companion who has their own home and car. Now I have traveled this whole planet and still do, and I know most people often come face–to–face with someone of the opposite sex to whom they become instantly attracted. Sometimes it's the smile, a look that lasts a little longer than normal, or the tactile hand that touches yours. The point I am trying to make is that it's fairly easy to form a relationship, especially if both people are attracted to each other.

Living and working with people from a poor area is not as bad as it sounds. The locals are mainly friendly and you will be fed, but don't expect any wages for working.

In my own experiences, most of the people I know who have gone to live in another country are single men, and those that survived best are the ones who got involved with a local girl. Providing she is willing, not already married, or with a jealous boyfriend, it is a fairly good path to take. You will learn the language quicker, be taught the various customs, and understand the culture of the local society. To some degree, you will also be getting a small amount of protection in the way of an early warning should you step out of line in the eyes of the locals. A girlfriend can smooth things over and explain to you what you did wrong. Friendship within the local community is a vital part of fitting in and building your identity protection.

Dating Websites

While this might seem to be an odd subject to put in a book on how to disappear, it is, in my opinion, a vital part of your disguise and normality. Meeting someone new in a strange country is not as difficult as it might seem. True, there are some countries where society does not allow women the social freedom we are used to in the West and others where many marriages are arranged. Nonetheless, there are also many countries—including, interestingly some where the main religion is Islam or Hindu—where dating websites are operating with vigor.

The only thing you need to remember is that you will be making your image and profile available to the world. Make sure that, even if someone in your old country should stumble across the dating website and see an image of you, they do not recognize you. However, be warned: a skip tracer who has gotten wind that you may have gone to Brazil will in all likelihood be checking out the dating websites to try and find you.

Dating websites have got to be the simplest way of finding a new partner in the country you intend to settle in once you have disappeared. Selecting a dating website in that country should be part of your disappearance plan. If you are properly organized, you may have already started on a particular website and found a partner who is waiting for you to arrive. The great thing about dating websites is that you get to see a picture and some background information on the person. By selecting a list, you can identify those who may have their own home, are employed, have children, and those who are divorced, etc. While this might seem trivial, believe me, when you arrive at your country of destination and you are met at the airport by a beautiful woman who has her own home and car, your new life will get off to a much better start. Don't forget, most dating sites accommodate just about every sexual orientation.

There are, for the person who wants to disappear, a few basic rules that govern the type of partner you might wish to meet and settle with. Your first relationship may not last more than a few months, but at least you will be settled by then. As for choosing someone, we are all different but here is my own list, should I ever try to disappear.

- **Age:** Always choose someone of around thirty-five to forty, or within five years of your own age; slightly older partners will generally treat you better. Many younger women will go for older men for this very reason.

- **Looks:** Looks are not everything, the inner person is more important.
- **Working:** If a person is working, they will not be a drain on your resources. You should not take advantage of their earnings.
- **Children:** If a partner has children, this could add unnecessary difficulties (I personally love children), but remember this is cover for your disappearance.
- **Own Home:** This is a real bonus, as it will generally mean you will have somewhere to stay when you arrive. If you do move in, you will increase your chances of staying if you offer to pay some rent, which is still a lot cheaper than living in a hotel.
- **Own Car:** Again, a bonus, as it means you can get from place to place and learn your way around.
- **Also Check:** Health? Hobbies? Things in common? Drinking or smoking habits?

There are lots of beautiful people living in Brazil, many of whom can be found on dating websites.

Example from a Dating Website

Marcia Maria

Age	37
Marital Status	Single
Occupation	Admin/office worker
Children	1
Height	5 ft 2 in (158 cm)
Weight	115 lb (52 kg)
Hair Color	Dark brown
Eye Color	Brown
Religion	Catholic
Star Sign	Virgo
Smoking	Never
Drinking	Never
Languages Spoken	English, Portuguese, Basic Spanish
Nationality	Brazil
Residence	Brazil
City/Region	Brasilia Distrito Federal
Likes	cinema, beach, swimming, reading, restaurant, church
Wants	Marriage
Partner Sought:	honest, romantic, sincere, loving, generous, aged 35–58

Other Ways to Find a Partner

Once you have arrived and settled into a good budget hotel (not too sleazy, as you might need to impress your partner), your first task will be to take a tour of the area. Get a map, walk around, take the local transportation, and get to know the place—relax and enjoy yourself. The one thing you will need to avoid is going to expat communities; if you're an American, stay away from the American expat community as a matter of security. However, if you do hear someone speaking your language and they are obviously not from your country of origin (example: an American hearing a Dutchman speaking English), then you might try and open a conversation. The benefit of someone who has lived (and possibly worked) in the country for several years is invaluable. Over a few beers, they may happily tell you their brief life story, places to go and not go, and the good and bad points about the country.

Once you have gotten to know the place a little, you should try looking for a partner or a job. Believe it or not, there are many foreign countries that welcome the benefits of, for example, an engineer from overseas. The pay might not be great, but it's a start and should provide enough income to live on. Getting a job will really depend on what your skills are and what skills are needed.

Author's Note: I do consultancy work for a military company, and some time ago they employed a young and very talented graphic artist. One day out of the blue, he resigned and went to live in Indonesia. For the first few weeks he trawled the bar and expat haunts, but one night he got talking to a guy who put him in touch with a local firm that immediately gave him a job as a graphic designer. I met him on one of my visits to Jakarta; he was happy with his new life and in a relationship with a secretary from the firm.

Supermarket meeting has to be tried just to show how simple it is.

Supermarket Chat Up

I mention supermarket or store chat ups because it has worked for me for more years than I can remember. When I was younger, my favorite way of getting a girl when I was overseas was to start a conversation in a supermarket. I once went to purchase some underpants in Singapore and ended up dating the store assistant. Another time in the same city, I had several dates with a female taxi driver; both women were extremely pleasant, and if I had stayed longer, the relationship with either could have blossomed.

Many people think that Asia is just full of bars where foreigners go for sex; this could not be further from the truth. Most Asian countries have a booming economy, a decent standard of living, and a social life that, in many cases, is far better than what we currently enjoy. The hotels are some of the best in the world, and their supermarkets and stores are a match for any of those in the West.

Do you know why a supermarket is one of the best places to meet women? It makes no difference if it's your local convenience store or a major supermarket. Women are usually at ease in a supermarket and don't have their guard up like they do in a nightclub or other regular pickup spot. You will be amazed at just how easy it is to pick up women in a supermarket, and and not just in Asia. It works equally as well in Britain and America. In addition to people shopping, many of the check out people are single mothers or divorced. The secret is to look clean and fresh and smell nice. Dress casually and smile—look as if you own the world. If you go in at around 6:00 p.m., you could try dressing in a suit and tie as if you were returning from a day in the office. This gives the impression that you are employed and financially sound, which is a big attraction for some. There are several ways to start a conversation; bumping into their cart is the easiest. Apologize and then ask if they know where you can find the wine department or the meat counter.

Don't be too disappointed if your first effort fails miserably; have several goes and you will get lucky. When you do manage to get a reply, smile and say something like "Have I seen you in here before?" If that works, tell them your name and keep it going. While you are doing this look casually in their trolley, as this will tell you a lot about their status, i.e. meals for one are a dead give away while large boxes of washing detergent may indicate she is married with children.

Author's Note: A British man known as "Fast Eddie" worked for a money collection service in the UK called Securicor. One day he decided to remove the contents of his collection and make off into the sunset. He actually finished up in America, taking the bulk of the £1.2 million with him. This happened in 1993 when Eddie was driving a security pick-up van for the company. His colleague had just retrieved the money from Lloyds Bank in Hamilton Road, Felixstowe, and upon exiting the bank, placed it in the hatch. As he finished doing so, Eddie just drove off. The van was later found abandoned down at the seafront, with Eddie and the money nowhere to be found.

While chatting up another shopper in the store requires you to start a conversation, sales personnel and check-out staff are required to talk to you.

It looked like Eddie had planned the perfect robbery, as he covered his tracks well after his disappearance. He switched cars twice in quick succession, knowing the security van would be tracked and its location quickly found. He was soon on a flight to Boston, where his common law wife, Deborah Brett, was waiting for him, along with their three-year-old son, Lee. Despite a £100,000 reward being offered, Eddie disappeared without a trace, having executed the perfect crime.

For several years they kept on the move, going from state to state and eventually ending up in Ozark, Missouri. They used a wide variety of aliases and always rented their home, usually in a modest neighborhood where they adopted the lifestyle of a normal, hard-working American family. Eddie, who was adept with electronics, took to working as a cable technician, while Deborah played the happy housewife. All went well for almost twenty years, until Eddie's son started to boast about his father's secret life. Lee had grown up with the knowledge of

"Fast Eddie" robbed his own security van in the UK and made it successfully to America. Almost twenty years later, he was caught because he did not keep his mouth shut.

what his father had done and how they came to be in America. He had boasted to several of his girlfriends; one in particular, Jessica, whom he had gotten pregnant, married, and then left, knew the whole story.

It took some persuading to convince the local police that the story she was telling was true and not just some revenge on the husband who had deserted her. Eventually, they contacted the FBI, who in turn contacted Suffolk Police in the UK. Despite the change of facial features over the past nineteen years, there was a resemblance and eventually Eddie was arrested by the FBI.

The moral here is to never tell anyone—even your children—what you have done in the past. Eddie had disappeared with a clean getaway (although he would forever be looking over his shoulder). He had managed to get himself an American Social Security number and employment, but in the end, boasting had given him away. He finished up in court and was sentenced to five years in prison.

Finding Employment Overseas

Unless you are one of the privileged few with a lot of money, the chances are that, like the majority of us, you will have to work. This rule will still apply once you have disappeared. Looking for work in a foreign country is not as difficult as one might think, but you will need to check the rules for the individual country in which you seek to disappear. Most countries welcome skilled and professional labor, but it is also quite easy to find general work on a temporary work permit. It is also possible to work without any permits, but I would advise against it. Many people just enter the country and find work with only a visitor visa, but in many cases this is illegal and you are open to exploitation.

It's up to the individual to check with the country of final destination, but there are many routes for employment open. I believe that if you are a citizen of a country that is part of the Commonwealth (that includes Canada, the UK, South Africa, Australia, and New Zealand) and you are under twenty-five years old, then you are eligible for a visa that will allow you to travel and work in the other Commonwealth countries. This may have changed or now have restrictions, so I suggest you check with your local embassy or consulate.

Switzerland, for example, has a high migration rate—foreigners make up more than a fifth of the Swiss population. The country is divided into twenty-six cantons (regions), and each is responsible for registering foreign workers. Although all cantons operate under the same federal law, each canton has some autonomy over immigration into the region. Anyone who works during their stay in Switzerland, or who remains in Switzerland for longer than three months, requires a residence permit issued by the Cantonal Migration Offices. There are short-term residence permits (less than one year), annual residence permits (limited), and permanent resident permits (unlimited) available.

Obtaining a work permit differs according to your place of origin. Switzerland has a dual system for the admission of foreign workers, with priority given to individuals from EU/EFTA countries. For employed nationals from EU/EFTA states, the Agreement on the Free Movement of Persons applies, leading to a straightforward permit process that is not subject to quotas. So if you're a member of the European Community, you can easily get a job.

The sort of work you look for will again depend on your own personal skills; people with a profession or trade will always find lots of opportunities. These who are unskilled can always find less attractive but enjoyable work if they look in the right place. Here are a few possibilities.

Construction Industry

Construction work is generally hard, but the pay is much better than, say, a bartender. The world has lots of places where construction jobs abound. While Europe is currently suffering from the results of over-construction, the Middle and Far East are booming.

Working on a cruise ship is a great life, but you will need skills and it is best if you are young.

Bartending and Hotel Work

Bartending and hotel work can be fun and very rewarding. If you plan to do this type of work overseas, then you are going to have to speak the local language. My advice is to find a good bar or a known hotel chain.

Teaching Your Native Language

This is an easy one. Get yourself on a TEFL course and teach your language to others. Likewise, if you already have knowledge of foreign languages, see if you can use it to your best advantage. Did you know that the American military paid interpreters about $800 per day to go out with US troops and act as interpreters in Afghanistan?

Sports

If you excel at a particular sport, you might try using this skill to find employment. There are many places where sports skills can find you work, e.g. a ski instructor in Switzerland, aerobics instructor on board a cruise ship, or athletics coach in a foreign school.

General

If you are not very good at anything, you can always find casual work caring for others, cleaning houses, or waiting tables in a restaurant.

Living and Working in a Foreign Country

Living in a new culture can be exciting, challenging, frustrating, and sometimes downright frightening. It is one thing to visit a country while on holiday and quite another to actually live there. Vacationers see the better side of life, live in smart hotels, and lay around on the beach. If you live in a country, you will have to work (unless you have sufficient funds) and live in normal accommodations. By normal, I mean housing similar to what the local people have. Now this can differ from place to place; it might be a small apartment in a very crowded suburb of Bangkok, or it could be a simple homemade log cabin in the jungles of South America. In either case, you will need to know in advance what type of living accommodation you will be in for once you reach your destination. Before departure, learn about the country's history, social customs, any unwritten rules, the current political structure, and above all, its main religion. Read up on the country's present day problems and current national issues.

You will need to know how to blend in with the neighbors, if you have any: shop as they shop, clean as they clean, and work as they work. Above all, you will need to fit into their society, adapt to their behavior, and become accustomed to their ways, habits, and culture.

You do not have to do this all at once; you can do it step by step. However, I would suggest that, while it is a good idea to make friends, it is not a good idea to get too involved too soon with the locals. That will come in due course—let them get used to you just as you get used to them.

Unless you are a well-traveled and seasoned adventurer, there is the possibility that you will become prone to culture shock. It is not uncommon in some poorer countries for children and adults alike to defecate and urinate in the open, although in my personal experience, most people do try to do their ablutions in private. Dogs may roam free and scavenge the trash, motorbikes and scooters might keep you awake at night, and then there are all the nasty creepy crawlies. For some who have never had to put up with such dreads, it can all come as bit of a culture shock.

The answer is not to give in and to follow the locals in everything they do. Make sure your personal accommodation is safe and free of any bugs, rats, or cats, and make provisions to keep them out. Check your bed every night so you can have a safe night's sleep, and wear ear defenders if the noise is too loud. Locate the best place to wash and do your ablutions; if you are shy, have some toilet provision within your own living space. The same goes for cooking. While I love eating street food, if you find it not to your liking, make plans to cook for yourself.

Living in a new country requires understanding. There is no point in saying, "In America we do it this way, which is better." Each country has its own ways—live with it.

At the end of the day, if the area you are living in becomes too much, simply move. It might take several moves before you find a place that suits your needs, though, so be patient. Take into account that when you first arrive in a new country it is exciting, but this will wear off and you may get a little depressed or frustrated. One of the first things you will start to do is compare the way of life in your new country to that of the one you left behind. North Americans and Europeans live a good life in contrast to many places in the world, but once again, in my experience, most people—no matter how poor—have respect, are clean, and obey the law.

True, there are some places that are downright dangerous, and the people—mainly the male population—are extremely aggressive. If you did your homework and stuck to your disappearance plan, this should not affect you, as you will have chosen to live among good people. You will simply need to give your new life time and gradually adjust to your new environment. It might take time, but you will eventually become familiar with the new culture, and the locals will treat you as one of their own. Your dress and attitude will moderate, and the foreigner in you will lessen.

Author's Note: Take time to think about your old habits. If someone comes looking for you and you maintain your old habits, they will quickly pick you out from the rest of the locals. So be careful of reading an American magazine or newspaper, smoking US cigarettes, ordering US labeled beer, and so on. As previously mentioned, learn the local language, read the local papers, drink the local beer, and smoke the local cigarettes.

Once you have completed your disappearing act, there is no going back unless you give up. Stay the course, find yourself a good safe niche within the local community, adjust, and make it enjoyable. Remember, when visiting and mingling with foreign cultures, keep in mind these tips, which may help ease you into cultural adjustment:

- Do your homework before you leave; check out the dos and don'ts of the country to which you intend to go.
- Be humble, greet and talk to local people, and be friendly.
- Listen and observe, learn the local ways quickly.
- Never discuss religion and never convert unless you genuinely want to.
- Do not expect to find things the same as back home, and do not harp on about the differences or how good the system is that you have just left.
- Keep an open mind and enjoy all the new experiences.
- Be a good ambassador for your country and your race.
- Remember, no matter how good the people, there will always be one who dislikes you.
- Remember you are a guest in their country; treat your hosts— rich or poor—with respect.
- No one is inferior, cultures and people are just different.
- Never make a promise you will not keep.

Summary

Once you have disappeared, stick to your original plan or the past will catch up with you. Stick to your cover story and be the person people perceive you to be. If you have moved overseas and have legally found employment, work within the laws of that country.

No one likes to be alone, so once you are settled, see if you can find yourself a partner. This will help you integrate with your new country in many ways. It will also add to your protection should people come looking for you.

If you do decide to simply walk away from your troubles and become a hobo, take my advice and do a little preparation; it's easy to degenerate into a real tramp. Keeping yourself clean and tidy will always help your case; people will look favorably on you and offer you work, or provide you with a meal and a bed.